# Numerical Methods for Convex Multistage Stochastic Optimization

## Other titles in Foundations and Trends® in Optimization

*A Tutorial on Hadamard Semidifferentials*
Kenneth Lange
ISBN: 978-1-63828-348-5

*Massively Parallel Computation: Algorithms and Applications*
Sungjin Im, Ravi Kumar, Silvio Lattanzi, Benjamin Moseley and Sergei
Vassilvitskii
ISBN: 978-1-63828-216-7

*Acceleration Methods*
Alexandre d'Aspremont, Damien Scieur and Adrien Taylor
ISBN: 978-1-68083-928-9

*Atomic Decomposition via Polar Alignment: The Geometry of Structured
Optimization*
Zhenan Fan, Halyun Jeong, Yifan Sun and Michael P. Friedlander
ISBN: 978-1-68083-742-1

*Optimization Methods for Financial Index Tracking: From Theory to
Practice*
Konstantinos Benidis, Yiyong Feng and Daniel P. Palomar
ISBN: 978-1-68083-464-2

# Numerical Methods for Convex Multistage Stochastic Optimization

**Guanghui Lan**
Georgia Institute of Technology
george.lan@isye.gatech.edu

**Alexander Shapiro**
Georgia Institute of Technology
ashapiro@isye.gatech.edu

the essence of knowledge
Boston — Delft

# Foundations and Trends® in Optimization

*Published, sold and distributed by:*
now Publishers Inc.
PO Box 1024
Hanover, MA 02339
United States
Tel. +1-781-985-4510
www.nowpublishers.com
sales@nowpublishers.com

*Outside North America:*
now Publishers Inc.
PO Box 179
2600 AD Delft
The Netherlands
Tel. +31-6-51115274

The preferred citation for this publication is

G. Lan and A. Shapiro. *Numerical Methods for Convex Multistage Stochastic Optimization.* Foundations and Trends® in Optimization, vol. 6, no. 2, pp. 63–144, 2024.

ISBN: 978-1-63828-351-5
© 2024 G. Lan and A. Shapiro

# Foundations and Trends® in Optimization
## Volume 6, Issue 2, 2024
## Editorial Board

# Editorial Scope

Foundations and Trends® in Optimization publishes survey and tutorial articles in the following topics:

- algorithm design, analysis, and implementation (especially, on modern computing platforms

- models and modeling systems, new optimization formulations for practical problems

- applications of optimization in machine learning, statistics, and data analysis, signal and image processing, computational economics and finance, engineering design, scheduling and resource allocation, and other areas

## Information for Librarians

Foundations and Trends® in Optimization, 2024, Volume 6, 4 issues. ISSN paper version 2167-3888. ISSN online version 2167-3918. Also available as a combined paper and online subscription.

# Contents

1  Introduction                                                          3

2  Stochastic Programming                                                8

3  Stochastic Optimal Control                                           14

4  Risk Averse and Distributionally Robust Optimization                 21

5  Dynamic Cutting Plane Algorithms                                     27
   5.1  SDDP Algorithm for SP Problems . . . . . . . . . . . .          28
   5.2  Cutting Plane Algorithm for SOC Problems . . . . . . .          40
   5.3  SDDP Algorithms for Risk Averse Problems . . . . . . .          43

6  Computational Complexity of Cutting Plane Methods                    55
   6.1  Computational Complexity of SDDP for SP . . . . . . .           55
   6.2  Explorative Dual Dynamic Programming for SP . . . . .           58
   6.3  Complexity of SDDP and EDDP for SOC . . . . . . . .             60
   6.4  Complexity of SDDP and EDDP for Risk Averse Problems            61
   6.5  Complexity of SDDP and EDDP over an Infinite Horizon .          61

7  Dynamic Stochastic Approximation Algorithms                         63
   7.1  Extension of Stochastic Approximation . . . . . . . . . .       63
   7.2  Approximate Stochastic Subgradients . . . . . . . . . . .       65

7.3   The DSA Algorithm and its Convergence Properties . . . .   67

7.4   DSA for General Multistage Stochastic Optimization   . . .   69

7.5   DSA for SOC Problems . . . . . . . . . . . . . . . . . . .   71

7.6   Combined EDDP and DSA for Hierarchical Problems   . . .   71

**8   Conclusions**                                                 **73**

**Acknowledgements**                                                **75**

**References**                                                      **76**

# Numerical Methods for Convex Multistage Stochastic Optimization

Guanghui Lan and Alexander Shapiro

*Georgia Institute of Technology, USA; george.lan@isye.gatech.edu, ashapiro@isye.gatech.edu*

ABSTRACT

Optimization problems involving sequential decisions in a stochastic environment were studied in Stochastic Programming (SP), Stochastic Optimal Control (SOC) and Markov Decision Processes (MDP). In this monograph, we mainly concentrate on SP and SOC modeling approaches. In these frameworks, there are natural situations when the considered problems are convex. The classical approach to sequential optimization is based on dynamic programming. It has the problem of the so-called "curse of dimensionality", in that its computational complexity increases exponentially with respect to the dimension of state variables. Recent progress in solving convex multistage stochastic problems is based on cutting plane approximations of the cost-to-go (value) functions of dynamic programming equations. Cutting plane type algorithms in dynamical settings is one of the main topics of this monograph. We also discuss stochastic approximation type methods applied to multistage stochastic optimization problems. From the computational complexity point of view, these two types of methods seem to be complementary to each other. Cutting plane type methods can handle multistage problems with a large number of stages

Guanghui Lan and Alexander Shapiro (2024), "Numerical Methods for Convex Multistage Stochastic Optimization", Foundations and Trends® in Optimization: Vol. 6, No. 2, pp 63–144. DOI: 10.1561/2400000044.

but a relatively smaller number of state (decision) variables. On the other hand, stochastic approximation type methods can only deal with a small number of stages but a large number of decision variables.

**Keywords**: Stochastic programming, Stochastic optimal control, Markov decision process, Dynamic programming, Risk measures, Stochastic dual dynamic programming, Stochastic approximation method, Cutting plane algorithm.

**AMS subject classifications**: 65K05, 90C15, 90C39, 90C40.

# 1

## Introduction

Traditionally different communities of researchers dealt with optimization problems involving uncertainty, modeled in stochastic terms, using different terminology and modeling frameworks. In this respect we can point to the fields of Stochastic Programming (SP), Stochastic Optimal Control (SOC) and Markov Decision Processes (MDP). Historically the developments in SP on the one hand, and SOC and MDP on the other, went along different directions with different modeling frameworks and solution methods. SOC is an interesting model since it can not only be naturally written in the MDP terms, but also can be formulated in the SP framework. In this monograph we mainly concentrate on SP approaches, and often specialize them to SOC whenever possible to demonstrate some basic ideas that can potentially bridge these three communities.

In these modeling frameworks mentioned above, there exist some natural situations when the considered problems are *convex*. An optimization problem is said to be convex if both its objective function and feasible set are convex. It is well-known that convexity provides the main apparatus for the development of efficient numerical algorithms for continuous optimization [45], [46]. The main goal of this work is to

3

present some recent developments in numerical approaches to solve convex optimization problems involving sequential decision making. Note that we do not intend to give a comprehensive review of the subject with a complete list of references. *Rather the aim is to present a certain point of view about some recent developments in solving* convex *multistage stochastic programming problems.*

Stochastic Programming (SP) has a long history. Two stage stochastic programming (with recourse) was introduced in Dantzig [10] and Beale [2], and was intrinsically connected with linear programming. From the beginning SP aimed at numerical solutions. Until about twenty years ago, the modeling approach to two and multistage SP was predominately based on construction of scenarios represented by scenario trees. This approach allows one to formulate the so-called deterministic equivalent optimization problem with the number of decision variables more or less proportional to the number of scenarios. When the deterministic equivalent could be represented as a linear program, such problems were considered to be numerically solvable. Because of that, the topic of SP was often viewed as a large scale linear programming. Further discussion and development of this approach can be found in Birge [6] and references therein.

From the point of view of the scenarios construction approach there is not much difference between two stage and multistage SP. In both cases the numerical effort in solving the deterministic equivalent is more or less proportional to the number of generated scenarios. This view on SP started to change with developments of randomization methods and the sample complexity theory [68]. From the perspective of solving the deterministic equivalent problem, even two stage linear stochastic programs are computationally intractable; their computational complexity is #P-hard for a sufficiently high accuracy, implying that they are at least as hard as NP problems (cf., [14], [23]). On the other hand, under reasonable assumptions, the number of randomly generated scenarios (by Monte Carlo sampling techniques), which are required to solve two stage SP problems with accuracy $\varepsilon > 0$ and high probability is of order $O(\varepsilon^{-2})$, see [68, Section 5.3]. While randomization methods were reasonably successful in solving two stage problems, the situation is different as far as multistage SP is concerned. The number of scenarios needed to solve multistage SP problems grows exponentially with the increase of the number of stages, see [70] and [68, Section 5.8.2].

Classical approach to sequential optimization is based on *dynamic programming* [3]. Dynamic programming also has a long history and is at the heart of the SOC and MDP modeling. It has the problem of the so-called "Curse of Dimensionality", a term coined by Bellman [3]. Its computational complexity increases exponentially with respect to (w.r.t.) the dimension of state variables. There is a large literature intending to deal with this problem by using various approximations of dynamic programming equations (see [56] and the references therein). Most of these methods are heuristics and often do not give verifiable guarantees for the accuracy of obtained solutions. There exist some developments on approximate dynamic programming with performance guarantees, e.g. those based on fitted value/policy iteration [41]–[43] and policy gradient methods [30], [32]. However, these performance guarantees often depend on an unknown function approximation error associated with the expressiveness of a given function class used to approximate the cost-to-go (value) functions.

Recent progress in solving convex multistage SP problems is based on cutting plane approximations of the cost-to-go functions of dynamic programming equations. These methods allow to estimate the error of the computed solution. Cutting plane type algorithms in dynamical settings is one of the main topics of this work. In particular, Stochastic Dual Dynamic Programming (SDDP), an algorithm first introduced by Pereira and Pinto [47] that builds upon the nested decomposition algorithm of Birge [5], has been a popular cutting plane method for multistage SP. Its convergence properties have been extensively studied in the literature (see, e.g., [12], [19], [24], [31], [49], [65], [77]). In this monograph, we will discuss cutting plane algorithms in the frameworks of SP and SOC with a focus on their associated rate of convergence. Moreover, we will also present extensions of stochastic approximation (a.k.a. stochastic gradient descent) type methods [29], [44], [45], [57] for multistage stochastic optimization, referred to as dynamic stochastic approximation in [34]. From the computational complexity point of view, these two types of methods seem to be complimentary to each other in the following sense. Specifically, certain variants of cutting plane methods have a computational complexity that grows mildly (linearly or quadratically) w.r.t. the number of stages, but exponentially w.r.t. the dimension of decision variables. On the other hand, the computational complexity for dynamic stochastic approximation methods increases exponentially w.r.t. the number of stages, but only mildly depends on the dimension of decision variables. Therefore, cutting plane type methods can handle multistage problems with a large number of stages, but a relatively small number of state (decision) variables. On the other hand, stochastic approximation type methods can only deal with a small number of stages, but a large number of decision variables. These methods share the following common features: (a) both methods utilize the convex structure of the cost-to-go (value) functions of dynamic programming equations, (b) both methods do not require explicit discretization of the state space, (c) both methods guarantee the convergence to the global optimality, (d) rates of convergence for both methods have been established.

It is worth noting a few alternative numerical methods for solving convex multistage SP problems that will not be covered in detail in this monograph. Firstly, the progressive hedging algorithm by Rockafellar and Wets [58] is a well-known scenario-based decomposition method, which basically applies the alternating direction method of multipliers (ADMM) to handle linear non-anticipativity constraints in the randomly generated sample average approximation problem [68]. In fact, one can also apply other primal-dual first-order optimization methods to handle these linear constraints (see, e.g., Chapter 3 of [29]). However, the size of the decomposition problem, i.e., the number of decision variables and linear constraints, will grow exponentially with the number of stages. Hence, these methods can only be applied to problems with a small number of stages. In addition, different from SA method, these decomposition methods would require the scenario tree to be generated and saved in the computer memory. Secondly, some advanced cutting plane methods, e.g., those based on bundle level method [28], [36], [37], can be used for solving two-stage SP problems efficiently. However, their extensions to multistage SP appear to be nontrivial.

This monograph is organized as follows. SP and SOC models will be first discussed in Sections 2 and 3, respectively. In Section 4, we present

risk averse and distributionally robust SP and SOC models. Sections 5
and 6, respectively, are dedicated to cutting plane methods and their
rates of convergence. In Section 7, we review some recent progress on
SA methods for multistage stochastic optimization. This work concludes
with a brief summary and possible future research directions in Section 8.
Readers certainly do not need to strictly follow the above outline. For
example, beginners can skip the more technically involved discussion
of risk averse models in Section 4, and move directly to algorithmic
studies in their first pass through this work. It should be pointed out
that we attempt to cover the fundamental models (SP, SOC, and risk
aversion) in earlier sections, and discuss numerical methods in later
sections. However, we also cover some other models in later sections,
including infinite horizon models, periodic models, and hierarchical
models, since the development of these models was inspired by the
studies on numerical methods for multistage SP.

We use the following notation and terminology throughout the mono-
graph. For $a \in \mathbb{R}$ we denote $[a]_+ := \max\{0, a\}$. Unless stated otherwise
$\|\cdot\|$ denotes Euclidean norm in $\mathbb{R}^n$. By $\text{dist}(x, S) := \inf_{y \in S} \|x - y\|$ we
denote the distance from a point $x \in \mathbb{R}^n$ to a set $S \subset \mathbb{R}^n$. We write
$x^\top y$ or $\langle x, y \rangle$ for the scalar product $\sum_{i=1}^n x_i y_i$ of vectors $x, y \in \mathbb{R}^n$. It is
said that a set $S \subset \mathbb{R}^n$ is polyhedral if it can be represented by a finite
number of affine constraints, it is said that a function $f : \mathbb{R}^n \to \mathbb{R}$ is
polyhedral if it can be represented as maximum of a finite number of
affine functions. For a process $\xi_1, \xi_2, \ldots$, we denote by $\xi_{[t]} = (\xi_1, \ldots, \xi_t)$ its
history up to time $t$. By $\mathbb{E}_{|X}[\cdot]$ we denote the conditional expectation,
conditional on random variable (random vector) $X$. We use the same
notation $\xi_t$ viewed as a random vector or as a vector variable, the particular meaning will be clear from the context. For a probability space
$(\Omega, \mathcal{F}, \mathbb{P})$, by $L_p(\Omega, \mathcal{F}, \mathbb{P})$, $p \in [1, \infty)$, we denote the space of random
variables $Z : \Omega \to \mathbb{R}$ having finite $p$-th order moment, i.e., such that
$\int |Z|^p d\mathbb{P} < \infty$. Equipped with norm $\|Z\|_p := (\int |Z|^p d\mathbb{P})^{1/p}$, $L_p(\Omega, \mathcal{F}, \mathbb{P})$
becomes a Banach space. The dual of $\mathcal{Z} := L_p(\Omega, \mathcal{F}, \mathbb{P})$ is the space
$\mathcal{Z}^* = L_q(\Omega, \mathcal{F}, \mathbb{P})$ with $q \in (1, \infty]$ such that $1/p + 1/q = 1$.

# 2

---

# Stochastic Programming

---

In Stochastic Programming (SP) the $T$-stage optimization problem can be written as

$$\min \quad \mathbb{E}\left[\sum_{t=1}^{T} f_t(x_t, \xi_t)\right] \tag{2.1}$$

$$\text{s.t} \quad x_t \in \mathcal{X}_t, \quad B_t x_{t-1} + A_t x_t = b_t, \ t = 1, ..., T. \tag{2.2}$$

Here $f_t : \mathbb{R}^{n_t} \times \mathbb{R}^{d_t} \to \mathbb{R}$ are objective functions, $\xi_t \in \mathbb{R}^{d_t}$ are random vectors, $\mathcal{X}_t \subset \mathbb{R}^{n_t}$ are nonempty closed sets, $B_t = B_t(\xi_t)$, $A_t = A_t(\xi_t)$ and $b_t = b_t(\xi_t)$ are matrix and vector functions of $\xi_t$, $t = 1, ..., T$, with $B_1 = 0$. The optimization in (2.1) is performed over decision variables $x_t$, $t = 1, ..., T$, which are *functions* of the random process $\xi_t$, $t = 1, ..., T$. With some abuse of the notation we use the same notation $x_t$ to be a function of the random data process and vector in $\mathbb{R}^{n_t}$. The particular meaning will be clear from the context. We will elaborate on this in Remark 2.1 below.

Constraints (2.2) often represent some type of balance equations between successive stages, while the set $\mathcal{X}_t$ can be viewed as representing local constraints at time $t$. It is possible to consider balance equations in a more general form. Nevertheless, formulation (2.2) will be sufficient for our purpose of discussing convex problems. In particular, if the objective functions $f_t(x_t, \xi_t) := c_t^\top x_t$ are linear, with $c_t = c_t(\xi_t)$, and the sets $\mathcal{X}_t$ are polyhedral, then (2.1)-(2.2) becomes a linear multistage stochastic program.

The sequence of random vectors $\xi_1, ...,$ is viewed as a data process. Unless stated otherwise we make the following assumption throughout the work.

**Assumption 2.1.** Probability distribution of the random process $\xi_t$, $t = 1, ...,$ does not depend on our decisions, i.e., is independent of the chosen decision policy.

**Remark 2.1.** At every time period $t = 1, ...,$ we have information about (observed) realization $\xi_{[t]} = (\xi_1, ..., \xi_t)$ of the data process. That is, at time $t$ values $\xi_1, ..., \xi_t$, of the data process are known to the decision maker (see Figure 2.1 for decision epochs in SP). Naturally the decisions should be based on the information available at time of the decision and should not depend on the future unknown values $\xi_{t+1}, ..., \xi_T$; this is the so-called principle of *non-anticipativity*. That is, the decision variables $x_t = x_t(\xi_{[t]})$ in (2.1)-(2.2) are *functions* of the data process, and a sequence of such functions for every stage $t = 1, ..., T$, is called a *policy* or a decision rule. The optimization in (2.1)-(2.2) is performed over policies satisfying the feasibility constraints. The feasibility constraints of problem (2.1)-(2.2) should be satisfied with probability one, and the expectation is taken with respect to the probability distribution of the random vector $\xi_{[T]} = (\xi_1, ..., \xi_T)$. It is worthwhile to emphasize that it suffices to consider policies depending only on the history of the data process without history of the decisions, because of Assumption 2.1.

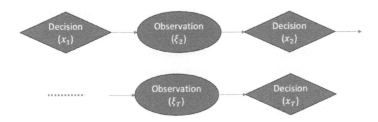

**Figure 2.1:** Decision epochs for SP.

The first stage decision $x_1$ is of the main interest and assumed to be *deterministic*, i.e., made before observing future realization of the data process. In that framework vector $\xi_1$ and the corresponding parameters $(A_1, b_1)$ are deterministic, i.e., their values are known at the time of first stage decision. Also the first stage objective function $f_1(x_1) = f_1(x_1, \xi_1)$ is a function of $x_1$ alone.                                                    □

We can write the following dynamic programming equations for (2.1)-(2.2) (e.g., [68, Section 3.1]). At the last stage the *cost-to-go (value)* function is

$$Q_T(x_{T-1}, \xi_T) = \inf_{x_T \in \mathcal{X}_T} \{f_T(x_T, \xi_T) : B_T x_{T-1} + A_T x_T = b_T\} \quad (2.3)$$

and then going backward in time for $t = T - 1, ..., 2$, the cost-to-go function is

$$Q_t(x_{t-1}, \xi_{[t]}) = \inf_{x_t \in \mathcal{X}_t} \left\{ f_t(x_t, \xi_t) + \mathbb{E}_{|\xi_{[t]}}[Q_{t+1}(x_t, \xi_{[t+1]})] : B_t x_{t-1} + A_t x_t = b_t \right\}.$$
$$(2.4)$$

The optimal value of (2.1)-(2.2) is given by the optimal value of the first stage problem

$$\min_{x_1 \in \mathcal{X}_1} f_1(x_1) + \mathbb{E}[Q_2(x_1, \xi_2)] \text{ s.t. } A_1 x_1 = b_1. \quad (2.5)$$

The optimization in (2.3)-(2.5) is over *nonrandom* variables $x_t \in \mathbb{R}^{n_t}$ (recall that $B_t = B_t(\xi_t)$, $A_t = A_t(\xi_t)$ and $b_t = b_t(\xi_t)$ are functions of $\xi_t$). The corresponding optimal policy is defined by

$$\bar{x}_t \in \arg\min_{x_t \in \mathcal{X}_t} \left\{ f_t(x_t, \xi_t) + \mathbb{E}_{|\xi_{[t]}}[Q_{t+1}(x_t, \xi_{[t+1]})] : B_t \bar{x}_{t-1} + A_t x_t = b_t \right\},$$
$$(2.6)$$

$t = 1, ..., T$, with $B_t \bar{x}_{t-1}$ for $t = 1$, and the expectation term of $Q_{T+1}(\cdot, \cdot)$ at the last stage omitted. We emphasize that the conditional expectation $\mathbb{E}_{|\xi_{[t]}}$ is a function of $\xi_{[t]}$, and hence the minimizer in the right hand side of (2.6) is a function of $\xi_{[t]}$. Therefore equation (2.6) defines a policy $\bar{x}_t = \bar{x}_t(\xi_{[t]})$, $t = 1, ..., T$. Of main interest is the first stage solution given by the optimal solution $\bar{x}_1$ of the first stage problem (2.5). As it was pointed before, the first stage decision $x_1$ is deterministic, made before observing realizations of the random process $\xi_t$, $t \geq 2$.

Since $\bar{x}_{t-1}$ is a function of $\xi_{[t-1]}$, the cost-to-go function $Q_t(\bar{x}_{t-1}, \xi_{[t]})$ can be considered as a function of $\xi_{[t]}$. It represents the optimal value of the problem,

$$\min \mathbb{E}\left[\sum_{\tau=t+1}^{T} f_\tau(x_\tau, \xi_\tau)\right] \text{ s.t. } x_\tau \in \mathcal{X}_\tau,\ B_\tau x_{\tau-1} + A_\tau x_\tau = b_\tau,\ \tau = t+1, ..., T,$$

(2.7)

conditional on realization $\xi_{[t]}$ of the data process. The optimization in (2.7) over $x_\tau = x_\tau(\xi_{[\tau]})$, $\tau = t+1, ..., T$, is often referred to as the dynamic programming (Bellman) principle of optimality. This is also the motivation for the name *cost-to-go* function. Another name for the cost-to-go functions is *value functions*.

**Remark 2.2.** Let us recall the following simple result. Consider an extended real valued function $\phi : \mathbb{R}^n \times \mathbb{R}^m \to \mathbb{R} \cup \{+\infty\}$. We have that if $\phi(x, y)$ is a convex function of $(x, y) \in \mathbb{R}^n \times \mathbb{R}^m$, then the min-value function $\varphi(x) := \inf_{y \in \mathbb{R}^m} \phi(x, y)$ is also convex. This can be applied to establish convexity of the cost-to-go functions. That is, suppose that the function $f_T(x_T, \xi_T)$ is convex in $x_T$ and the set $\mathcal{X}_T$ is convex. Then the extended real valued function taking value $f_T(x_T, \xi_T)$ if $x_T \in \mathcal{X}_T$ and $B_T x_{T-1} + A_T x_T = b_T$, and value $+\infty$ otherwise, is convex. Since the cost-to-go function $Q_T(x_{T-1}, \xi_T)$ can be viewed as the min-function of this extended real valued function, it follows that it is convex in $x_{T-1}$. Also note that if the cost-to-go function $Q_{t+1}(x_t, \xi_{[t+1]})$ is convex in $x_t$, then its expected value in (2.4) is also convex. Consequently by induction going backward in time it is not difficult to show the following convexity property of the cost-to-go functions.

**Proposition 2.1.** If the objective functions $f_t(x_t, \xi_t)$ are convex in $x_t$ and the sets $\mathcal{X}_t$ are convex, $t = 1, ..., T$, then the cost-to-go functions $Q_t(x_{t-1}, \xi_{[t]})$, $t = 2, ..., T$, are convex in $x_{t-1}$ and the first stage problem (2.5) is convex. In particular such convexity follows for *linear* multistage stochastic programs.

Such convexity property is crucial for development of efficient numerical algorithms. It is worthwhile to note that Assumption 2.1 is essential here. Dependence of the distribution of the random data on decisions typically destroys convexity even in the linear case. □

The dynamic programming equations reduce the problem from optimization over policies, i.e., over infinite dimensional functional spaces, to evaluation of the cost-to-go functions $Q_t(x_{t-1}, \xi_{[t]})$ of finite dimensional vectors. However, dependence of the cost-to-go functions on the whole history of the data process makes it numerically unmanageable with increase in the number of stages. The situation is simplified dramatically if we make the following assumption of *stagewise independence*.

**Assumption 2.2** (Stagewise independence). Probability distribution of $\xi_{t+1}$ is independent of $\xi_{[t]}$ for $t \geq 1$.

Under Assumption 2.2 of stagewise independence, the conditional expectations in dynamic programming equations (2.4) become the corresponding unconditional expectations. It is straightforward to show then by induction, going backward in time, that the dynamic programming equations (2.4) can be written as

$$Q_t(x_{t-1}, \xi_t) = \inf_{x_t \in \mathcal{X}_t} \{ f_t(x_t, \xi_t) + \mathcal{Q}_{t+1}(x_t) : B_t x_{t-1} + A_t x_t = b_t \},$$
$$(2.8)$$

where

$$\mathcal{Q}_{t+1}(x_t) := \mathbb{E}[Q_{t+1}(x_t, \xi_{t+1})], \qquad (2.9)$$

with the expectation in (2.9) taken with respect to the marginal distribution of $\xi_{t+1}$. In that setting one needs to keep track of the (expectation) cost-to-go functions $\mathcal{Q}_{t+1}(x_t)$ only. The optimal policy is defined by the equations

$$\bar{x}_t \in \arg\min_{x_t \in \mathcal{X}_t} \{ f_t(x_t, \xi_t) + \mathcal{Q}_{t+1}(x_t) : B_t \bar{x}_{t-1} + A_t x_t = b_t \}. \quad (2.10)$$

Note that here the optimal policy values can be computed iteratively, starting from optimal solution $\bar{x}_1$ of the first stage problem, and then by sequentially computing a minimizer in the right hand side of (2.10) going forward in time for $t = 2, ..., T$. Of course, this procedure requires availability of the cost-to-go functions $\mathcal{Q}_{t+1}(x_t)$. Observe that even under the stagewise independence assumption, the possible number of scenarios which corresponds to the number of paths from the root node to the leaf nodes in Figure 2.2, still grows exponentially w.r.t. $T$.

It is worthwhile to note that here the optimal policy decision $\bar{x}_t$ can be viewed as a function of $\bar{x}_{t-1}$ and $\xi_t$, for $t = 2, ..., T$. Of course,

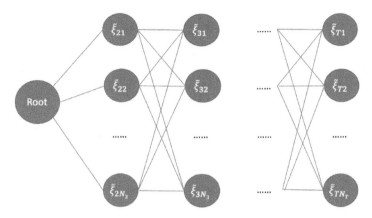

**Figure 2.2:** Representation of a scenario tree under stagewise independence assumption ($\tilde{\xi}_{ti}$, $i = 1, \ldots, N_t$, are random samples of $\xi_t$, $t = 2, \ldots, T$).

$\bar{x}_{t-1}$ is a function of $\bar{x}_{t-2}$ and $\xi_{t-1}$, and so on. So eventually $\bar{x}_t$ can be represented as a function of the history $\xi_{[t]}$ of the data process. Assumption 2.1 was crucial for this conclusion, since then we need to consider only the history of the data process rather than also the history of our decisions.

**Remark 2.3.** In the SP setting, the process $\xi_t$ is viewed as a data process, and the assumption of stagewise independence could be unrealistic. In order to model stagewise dependence we assume the following Markovian property of the data process: the conditional distribution of $\xi_{t+1}$ given $\xi_{[t]}$ is the same as the conditional distribution of $\xi_{t+1}$ given $\xi_t$. In such Markovian setting the dynamic programming equations (2.4) become

$$Q_t(x_{t-1}, \xi_t) = \inf_{x_t \in \mathcal{X}_t} \left\{ f_t(x_t, \xi_t) + \mathbb{E}_{|\xi_t}[Q_{t+1}(x_t, \xi_{t+1})] : B_t x_{t-1} + A_t x_t = b_t \right\}.$$
(2.11)

That is, compared with the stagewise independent case, in the Markovian setting the expected cost-to-go function

$$\mathcal{Q}_{t+1}(x_t, \xi_t) = \mathbb{E}_{|\xi_t}[Q_{t+1}(x_t, \xi_{t+1})]$$
(2.12)

also depends on $\xi_t$. □

# 3

## Stochastic Optimal Control

Consider the classical Stochastic Optimal Control (SOC) (discrete time, finite horizon) model (e.g., Bertsekas and Shreve [4]):

$$\min_{\pi} \ \mathbb{E}^{\pi} \left[ \sum_{t=1}^{T} c_t(x_t, u_t, \xi_t) + c_{T+1}(x_{T+1}) \right] \tag{3.1}$$

$$\text{s.t.} \quad u_t \in \mathcal{U}_t(x_t), \ t = 1, ..., T, \tag{3.2}$$

$$x_{t+1} = F_t(x_t, u_t, \xi_t), \ t = 1, ..., T. \tag{3.3}$$

Variables $x_t \in \mathbb{R}^{n_t}$, $t = 1, ..., T+1$, represent the state of the system, $u_t \in \mathbb{R}^{m_t}$, $t = 1, ..., T$, are controls, $\xi_t \in \mathbb{R}^{d_t}$, $t = 1, ..., T$, are random vectors, $c_t : \mathbb{R}^{n_t} \times \mathbb{R}^{m_t} \times \mathbb{R}^{d_t} \to \mathbb{R}$, $t = 1, ..., T$, are cost functions, $c_{T+1}(x_{T+1})$ is a final cost function, $F_t : \mathbb{R}^{n_t} \times \mathbb{R}^{m_t} \times \mathbb{R}^{d_t} \to \mathbb{R}^{n_{t+1}}$ are (measurable) mappings and $\mathcal{U}_t(x_t)$ is a (measurable) multifunction mapping $x_t \in \mathbb{R}^{n_t}$ to a subset of $\mathbb{R}^{m_t}$. Relation (3.3) is called the *state equation*. Values $x_1$ and $\xi_0$ are deterministic (initial conditions); it is also possible to view $x_1$ as random with a given distribution, but this is not essential for the following discussion (see Figure 3.1 for the decision epochs of SOC). The optimization in (3.1)-(3.3) is performed over policies $\pi$ determined by decisions $u_t$ and state variables $x_t$. This is emphasized in the notation $\mathbb{E}^{\pi}$, we are going to discuss this below.

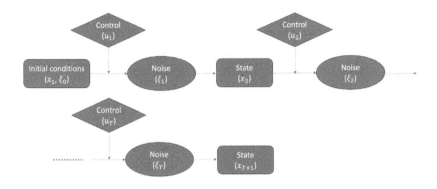

**Figure 3.1:** Decision epochs for SOC.

Unless stated otherwise we assume that probability distribution of the random process $\xi_t$ does not depend on our decisions (Assumption 2.1). We can consider (3.1)-(3.3) in the framework of SP if we view $y_t = (x_t, u_t)$, $t = 1, ..., T$, as decision variables. Suppose that the mapping $F_t(x_t, u_t, \xi_t)$ is affine, i.e.,

$$F_t(x_t, u_t, \xi_t) := A_t x_t + B_t u_t + b_t, \ t = 1, ..., T, \tag{3.4}$$

where $A_t = A_t(\xi_t)$, $B_t = B_t(\xi_t)$ and $b_t = b_t(\xi_t)$ are matrix and vector valued functions of $\xi_t$. The constraints $u_t \in \mathcal{U}_t(x_t)$ can be viewed as local constraints in the SP framework; in particular if $\mathcal{U}_t(x_t) \equiv \mathcal{U}_t$ are independent of $x_t$, then $\mathcal{U}_t$ can be viewed as a counterpart of the set $\mathcal{X}_t$ in (2.1)-(2.2). For the affine mapping of the form (3.4), the state equations (3.3) become

$$x_{t+1} - A_t x_t - B_t u_t = b_t, \ t = 1, ..., T. \tag{3.5}$$

These state equations are linear in $y_t = (x_t, u_t)$ and can be viewed as a particular case of the balance equations (2.2) of the stochastic program (2.1)-(2.2). Note, however, that here decision $u_t$ should be made before realization of $\xi_t$ becomes known, and $x_t$ and $u_t$ are functions of $\xi_{[t-1]}$ rather than $\xi_{[t]}$. We emphasize that $x_t$ and $u_t$ can be considered as functions of the history $\xi_{[t-1]}$ of the random process $\xi_t$, without dependence on the history of the decisions. As in the SP framework, this follows from the dynamic programming equations discussed below, and Assumption 2.1 is essential for that conclusion.

**Remark 3.1.** There is a time shift in equations (3.5) as compared with the SP equations (2.2). We emphasize that in both the SOC and SP approaches, the decisions are made based on information available at time of the decision; this is the principle of non-anticipativity. Because of the time shift, the dynamic programming equations for the SOC problem are slightly different from the respective equations in SP. More important is that in the SOC framework there is a clear separation between the state and control variables. This will have an important consequence for the respective dynamic programming equations which we consider next.                                                                                  □

Similar to SP, the dynamic programming equations for the SOC problem (3.1)-(3.3) can be written as follows. At the last stage, the *value function*

$$V_{T+1}(x_{T+1}) = c_{T+1}(x_{T+1})$$

and, going backward in time for $t = T, ..., 1$, the *value functions*

$$V_t(x_t, \xi_{[t-1]}) = \inf_{\substack{u_t \in \mathcal{U}_t(x_t) \\ x_{t+1} = F_t(x_t, u_t, \xi_t)}} \mathbb{E}_{|\xi_{[t-1]}} \left[ c_t(x_t, u_t, \xi_t) + V_{t+1}(x_{t+1}, \xi_{[t]}) \right].$$

$$(3.6)$$

The optimization in the right hand side of (3.6) is performed jointly in $u_t$ and $x_{t+1}$. The optimal value of the SOC problem (3.1)-(3.3) is given by the first stage value function $V_1(x_1)$, and can be viewed as a function of the initial conditions $x_1$. Note that because of the state equation (3.3), equations (3.6) can be written in the following equivalent form

$$V_t(x_t, \xi_{[t-1]}) = \inf_{u_t \in \mathcal{U}_t(x_t)} \mathbb{E}_{|\xi_{[t-1]}} \left[ c_t(x_t, u_t, \xi_t) + V_{t+1}(F_t(x_t, u_t, \xi_t), \xi_{[t]}) \right].$$

$$(3.7)$$

The corresponding optimal policy is defined by

$$(\bar{u}_t, \bar{x}_{t+1}) \in \arg\min_{\substack{u_t \in \mathcal{U}_t(\bar{x}_t) \\ x_{t+1} = F_t(\bar{x}_t, u_t, \xi_t)}} \mathbb{E}_{|\xi_{[t-1]}} \left[ c_t(\bar{x}_t, u_t, \xi_t) + V_{t+1}(x_{t+1}, \xi_{[t]}) \right].$$

$$(3.8)$$

Similar to SP, it follows that $\bar{u}_t$ can be considered as a function of $\xi_{[t-1]}$, and $\bar{x}_{t+1}$ as a function of $\xi_{[t]}$. That is, $(\bar{u}_t, \bar{x}_t)$ is a function of $\xi_{[t-1]}$, this was already mentioned in the previous paragraphs. Again Assumption 2.1 is essential for the above derivations and conclusions.

**Remark 3.2.** Let us discuss convexity of the value functions. Consider the graph of the multifunction $\mathcal{U}_t(\cdot)$:

$$\mathrm{Gr}(\mathcal{U}_t) := \{(x_t, u_t) : u_t \in \mathcal{U}_t(x_t),\ x_t \in \mathbb{R}^{n_t}\}. \tag{3.9}$$

Note that minimization $\min_{u_t \in \mathcal{U}_t(x_t)} \varphi(x_t, u_t)$ of a function $\varphi : \mathbb{R}^{n_t} \times \mathbb{R}^{m_t} \to \mathbb{R} \cup \{+\infty\}$ can equivalently be written as $\min_{u_t \in \mathbb{R}^{m_t}} \hat{\varphi}(x_t, u_t)$, where function $\hat{\varphi}(x_t, u_t)$ coincides with $\varphi(x_t, u_t)$ when $u_t \in \mathcal{U}_t(x_t)$, and is $+\infty$ otherwise. Of course, the constraint $u_t \in \mathcal{U}_t(x_t)$ is the same as $(x_t, u_t) \in \mathrm{Gr}(\mathcal{U}_t)$. If function $\varphi(x_t, u_t)$ is convex and the set $\mathrm{Gr}(\mathcal{U}_t)$ is convex, then the corresponding function $\hat{\varphi}(x_t, u_t)$ is convex. Consequently in that case the min-function $\min_{u_t \in \mathcal{U}_t(x_t)} \varphi(x_t, u_t)$ is a convex function of $x_t$. By applying this observation and using induction going backward in time, it is not difficult to show that value functions $V_t(x_t, \xi_{[t-1]})$ are convex in $x_t$, $t = 1, ..., T$, if the following assumption holds.

**Assumption 3.1.** The function $c_{T+1}(\cdot)$ is convex, and for every $t = 1, ..., T$, the cost function $c_t(x_t, u_t, \xi_t)$ is convex in $(x_t, u_t)$ for a.e. $\xi_t$, the mapping $F_t$ is affine of the form (3.4), the graph $\mathrm{Gr}(\mathcal{U}_t)$ is a convex subset of $\mathbb{R}^{n_t} \times \mathbb{R}^{m_t}$, $t = 1, ..., T$.

Note that, in particular, $\mathrm{Gr}(\mathcal{U}_t)$ is convex if $\mathcal{U}_t(x_t) \equiv \mathcal{U}_t$ is independent of $x_t$ and the set $\mathcal{U}_t$ is convex. When $\mathcal{U}_t(x_t)$ is dependent on $x_t$, we need some constructive way to represent it in order to ensure convexity of $\mathrm{Gr}(\mathcal{U}_t)$. For example suppose that the feasibility sets are defined by inequality constraints as:

$$\mathcal{U}_t(x_t) := \{u_t \in \mathbb{R}^{m_t} : g_t(x_t, u_t) \le 0\}, \tag{3.10}$$

where $g_t : \mathbb{R}^{n_t} \times \mathbb{R}^{m_t} \to \mathbb{R}^{\ell_t}$. Then $\mathrm{Gr}(\mathcal{U}_t) = \{(x_t, u_t) : g_t(x_t, u_t) \le 0\}$. Consequently if every component of mapping $g_t(x_t, u_t)$ is a convex function, then $\mathrm{Gr}(\mathcal{U}_t)$ is convex. □

In SOC framework the random process $\xi_1, ...,$ is often viewed as a noise or disturbances (note that here $\xi_1$ is also random). In such cases the assumption of stagewise independence (Assumption 2.2) is natural. In the stagewise independent case the dynamic programming equations (3.7) simplify with the value functions $V_t(x_t)$, $t = 1, ..., T$, being functions of the state variables only, that is

$$V_t(x_t) = \inf_{u_t \in \mathcal{U}_t(x_t)} \mathbb{E}\left[c_t(x_t, u_t, \xi_t) + V_{t+1}(F_t(x_t, u_t, \xi_t))\right], \qquad (3.11)$$

where the expectation is taken with respect to the marginal distribution of $\xi_t$. Consequently the optimal policy is defined by the optimal controls $\bar{u}_t = \bar{\pi}_t(x_t)$, where

$$\bar{\pi}_t(x_t) \in \arg\min_{u_t \in \mathcal{U}_t(x_t)} \mathbb{E}\left[c_t(x_t, u_t, \xi_t) + V_{t+1}(F_t(x_t, u_t, \xi_t))\right]. \qquad (3.12)$$

That is, in the stagewise independent case it is suffices to consider policies with controls of the form $u_t = \pi_t(x_t)$.

**Remark 3.3.** Suppose that the distribution of $\xi_t$ can depend on $(x_t, u_t)$. That is, the distribution of the random process is effected by our decisions and consequently Assumption 2.1 is not satisfied. Under the stagewise independence Assumption 2.2, the dynamic programming equations (3.11) still hold and the optimal policy is determined by (3.12). However in that case $\bar{u}_t$ depends on the history $(\bar{x}_1, \bar{u}_1, ..., \bar{u}_{t-1}, \bar{x}_t)$ of the decision process. This history depends on the decisions (on the considered policy) and cannot be represented in terms of the process $\xi_t$ alone. It also could be noted that in the stagewise independence case, SOC can be formulated as Markov Decision Processes (MDP) with transition probabilities $P_t(\cdot|x_t, u_t)$ determined by the state equations (3.3). We will not pursue the MDP formulation in the following discussion, and will rather deal with the SOC model instead.                                                □

As an example let us consider the classical inventory problem.

**Example 3.1** (Inventory problem). Consider the inventory problem (e.g., [76]),

$$\min_{u_t \geq 0} \quad \mathbb{E}\left[\sum_{t=1}^{T} c_t u_t + \psi_t(x_t + u_t, D_t)\right] \qquad (3.13)$$

$$\text{s.t.} \quad x_{t+1} = x_t + u_t - D_t, \ t = 1, ..., T, \qquad (3.14)$$

where $c_t, b_t, h_t \in \mathbb{R}$ are the ordering cost, backorder penalty cost and holding cost per unit, respectively, $x_t \in \mathbb{R}$ is the current inventory level,

$u_t \in \mathbb{R}$ is the order quantity, $D_t \in \mathbb{R}$ is the demand at time $t = 1, ..., T$, and the deficit-surplus cost function $\psi$ is given by

$$\psi_t(y_t, d_t) := b_t[d_t - y_t]_+ + h_t[y_t - d_t]_+.$$

In this format $x_t$ are state variables, $u_t$ are controls, $D_t$ are random disturbances, and $\mathcal{U}_t := \mathbb{R}_+$ is the feasibility set (independent of state variables). At time $t = 1$, the inventory level $x_1$ and demand value $D_0$ are known (initial conditions). Decision about $u_t$ at stage $t = 1, ..., T$, should be made before a realization of the demand $D_t$ becomes known. That is, at time $t$ history $D_{[t-1]}$ of the demand process is available, but future realizations $D_t, ..., D_T$ are not known.

By making change of variables $y_t = x_t + u_t$, (3.13)-(3.14) is often written in the following equivalent form

$$\min_{y_t \geq x_t} \quad \mathbb{E}\left[ \sum_{t=1}^{T} c_t(y_t - x_t) + \psi_t(y_t, D_t) \right] \tag{3.15}$$

$$\text{s.t.} \quad x_{t+1} = y_t - D_t, \; t = 1, ..., T. \tag{3.16}$$

(3.15)-(3.16) can be viewed as an SP problem in decision variables $y_t$ and $x_t$.

It is convenient to write dynamic programming equations for the inventory problem as defined in the form (3.15)-(3.16), although the initial form (3.13)-(3.14) can also be rewritten in terms of dynamic programming equations. As before we assume here that distribution of the demand process does not depend on our decisions. At stage $t = T, ..., 1$, the value functions satisfy the following equation

$$V_t(x_t, D_{[t-1]}) = \inf_{y_t \geq x_t} \left\{ c_t(y_t - x_t) + \mathbb{E}_{|D_{[t-1]}}\left[ \psi_t(y_t, D_t) + V_{t+1}\left( y_t - D_t, D_{[t]} \right) \right] \right\}, \tag{3.17}$$

with $V_{T+1}(\cdot) \equiv 0$. The objective functions $c_t(y_t - x_t) + \psi_t(y_t, D_t)$ are convex in $(x_t, y_t)$ and hence value functions $V_t(x_t, D_{[t-1]})$ are convex in $x_t$ (see Remark 3.2).

The optimal policy of how much to order at stage $t$ is given by $\bar{u}_t = \bar{y}_t - \bar{x}_t$, where

$$\bar{y}_t \in \arg\min_{y_t \geq \bar{x}_t} \left\{ c_t(y_t - \bar{x}_t) + \mathbb{E}_{|D_{[t-1]}}\left[ \psi_t(y_t, D_t) + V_{t+1}\left( y_t - D_t, D_{[t]} \right) \right] \right\}. \tag{3.18}$$

That is, starting with the initial value $\bar{x}_1 = x_1$, value $\bar{y}_1$ is computed as an optimal solution of the minimization problem in the right hand side of (3.18) for $t = 1$ (at the first stage, since $D_0$ is deterministic, the conditional expectation in (3.18) is just the unconditional expectation with respect to $D_1$). Consequently the next inventory level $\bar{x}_2 = \bar{y}_1 - D_1$ is computed given an observed realization of $D_1$. Then $\bar{y}_2$ is computed as an optimal solution of the minimization problem in the right hand side of (3.18) for $t = 2$, the inventory level is updated as $\bar{x}_3 = \bar{y}_2 - D_2$, and so forth going forward in time. This defines an optimal policy with the inventory level and order quantity at stage $t$ being functions of observed realization of the history $D_{[t-1]}$ of the demand process.

If the demand process $D_t$ is stagewise independent, then value functions $V_t(x_t)$ do not depend on the history of the demand process, and (3.17) becomes

$$V_t(x_t) = \inf_{y_t \geq x_t} \left\{ c_t(y_t - x_t) + \mathbb{E}\left[ \psi_t(y_t, D_t) + V_{t+1}\left( y_t - D_t \right) \right] \right\}, \quad (3.19)$$

with the expectations taken with respect to the marginal distributions of the demand process. By using convexity of the objective function it is possible to show that in the stagewise independence case the so-called basestock policy is optimal. That is, the optimal control (optimal order quantity) $\bar{u}_t$ is a function of state $x_t$ and is given by

$$\bar{u}_t = \max\{x_t, y_t^*\} - x_t = [y_t^* - x_t]_+, \quad (3.20)$$

where $y_t^*$ is the (unconstrained) minimizer

$$y_t^* \in \arg\min_{y_t} \left\{ c_t y_t + \mathbb{E}\left[ \psi_t(y_t, D_t) + V_{t+1}\left( y_t - D_t \right) \right] \right\}. \quad (3.21)$$

That is, if at stage $t$ the current inventory level $x_t$ is greater than or equal to $y_t^*$, then order nothing; otherwise order $y_t^* - x_t$. Of course, computation of the (critical) values $y_t^*$ is based on availability of value functions $V_t(\cdot)$.                                                                                                   $\square$

# 4

---

# Risk Averse and Distributionally Robust Optimization

---

In (2.1)-(2.2) and (3.1)-(3.3), the costs are supposed to be minimized on average. For a particular realization of the random cost its value can be quite different from the average (the expectation value). A risk-averse approach (as opposed to risk neutral formulation) is aimed at controlling variability of the random costs. In this section, we first review the risk averse approach and its relation to distributionally robust optimization in the static setting, and we then present possible ways to extend them to the dynamical setting and discuss their associated challenges. We show that such extensions can be greatly simplified under the stagewise independence assumption.

Formally in the risk averse approach, the expectation functional is replaced by a risk measure. Let $(\Omega, \mathcal{F}, \mathbb{P})$ be a probability space and $\mathcal{Z}$ be a linear space of measurable functions (random variables) $Z : \Omega \to \mathbb{R}$. Risk measure $\mathcal{R}$ is a functional assigning a number to random variable $Z \in \mathcal{Z}$, that is, $\mathcal{R} : \mathcal{Z} \to \mathbb{R}$. In the risk neutral case, $\mathcal{R}$ is the expectation operator, i.e., $\mathcal{R}(Z) := \mathbb{E}[Z] = \int_{\Omega} Z(\omega) d\mathbb{P}(\omega)$. In order for this expectation to be well defined we have to restrict the space of considered random variables; for the expectation it is natural to take the space of integrable functions,[1] i.e., $\mathcal{Z} := L_1(\Omega, \mathcal{F}, \mathbb{P})$.

It was suggested in Artzner *et al.* [1] that reasonable risk measures should satisfy the following axioms. For any random variables $Z, Z' \in \mathcal{Z}$ and numbers $a$ and $\lambda \geq 0$,

$$
\begin{aligned}
\mathcal{R}(Z + Z') &\leq \mathcal{R}(Z) + \mathcal{R}(Z') &&\text{(subadditivity)}, \\
Z \leq Z' &\Rightarrow \mathcal{R}(Z) \leq \mathcal{R}(Z') &&\text{(monotonicity)}, \\
\mathcal{R}(Z + a) &= \mathcal{R}(Z) + a &&\text{(translation equivariance)}, \\
\mathcal{R}(\lambda Z) &= \lambda \mathcal{R}(Z) &&\text{(positive homogeneity)}.
\end{aligned}
$$

Risk measures satisfying these axioms were called *coherent*. For a thorough discussion of coherent risk measures we can refer to [15], [68].

It is possible to show that if the space $\mathcal{Z}$ is a Banach lattice, in particular if $\mathcal{Z} = L_p(\Omega, \mathcal{F}, \mathbb{P})$, then a (real valued) coherent risk measure $\mathcal{R} : \mathcal{Z} \to \mathbb{R}$ is continuous in the norm topology of $\mathcal{Z}$ (cf., [63, Proposition 3.1]). It follows then by the Fenchel-Moreau theorem that $\mathcal{R}$ has the following dual representation

$$
\mathcal{R}(Z) = \sup_{\zeta \in \mathfrak{A}} \int_\Omega Z(\omega)\zeta(\omega)d\mathbb{P}(\omega), \quad Z \in \mathcal{Z}, \tag{4.1}
$$

where $\mathfrak{A}$ is a convex weakly* compact subset of $\mathcal{Z}^*$. Moreover, the axioms of coherent risk measures imply that every $\zeta \in \mathfrak{A}$ is a density function, i.e., almost surely $\zeta(\omega) \geq 0$ and $\int_\Omega \zeta(\omega)d\mathbb{P}(\omega) = 1$ (cf., [63, Theorem 2.2]).

An important example of coherent risk measure is the Average Value-at-Risk:

$$
\begin{aligned}
\mathsf{AV@R}_\alpha(Z) &:= (1-\alpha)^{-1} \int_\alpha^1 F_Z^{-1}(\tau)d\tau \tag{4.2} \\
&= \inf_{\theta \in \mathbb{R}} \left\{ \theta + (1-\alpha)^{-1}\mathbb{E}[Z - \theta]_+ \right\}, \tag{4.3}
\end{aligned}
$$

where $\alpha \in [0,1)$ is a given risk level, and $F_Z(z) := \mathbb{P}(Z \leq z)$ is the cumulative distribution function (cdf) of $Z$ and $F_Z^{-1}(\tau) := \inf\{z : F_Z(z) \geq \tau\}$ is the respective quantile. (4.2) suggests that $\mathsf{AV@R}_\alpha(Z)$

---

[1] $L_p(\Omega, \mathcal{F}, \mathbb{P})$, $p \in [1, \infty)$, is denoted as the space of random variables $Z$ with finite $p$-th order moment, i.e., $\int_\Omega |Z|^p d\mathbb{P} < \infty$.

penalizes the upper quantiles of the distribution of random variable $Z$. It is natural in this example to use the space $\mathcal{Z} = L_1(\Omega, \mathcal{F}, \mathbb{P})$ and its dual $\mathcal{Z}^* = L_\infty(\Omega, \mathcal{F}, \mathbb{P})$. In various equivalent forms this risk measure was introduced in different contexts by different authors under different names, such as Expected Shortfall, Expected Tail Loss, Conditional Value-at-Risk; variational form (4.3) was suggested in [48], [60]. In the dual form, it has representation (4.1) with

$$\mathfrak{A} = \left\{ \zeta \in \mathcal{Z}^* : 0 \le \zeta \le (1-\alpha)^{-1}, \ \int_\Omega \zeta d\mathbb{P} = 1 \right\}.$$

Representation (4.1) can be viewed from the distributionally robust point of view. Recall that if $P$ is a probability measure on $(\Omega, \mathcal{F})$ which is absolutely continuous with respect to $\mathbb{P}$, then by the Radon - Nikodym Theorem it has density $\zeta = dP/d\mathbb{P}$. Consider the set $\mathfrak{M}$ of probability measures $P$, absolutely continuous with respect to $\mathbb{P}$, defined as $\mathfrak{M} := \{P : dP/d\mathbb{P} \in \mathfrak{A}\}$. Then (4.1) of functional $\mathcal{R}$ can be written as[2]

$$\mathcal{R}(Z) = \sup_{P \in \mathfrak{M}} \mathbb{E}_P[Z]. \tag{4.4}$$

In the Distributionally Robust Optimization (DRO) literature, it is argued that the "true" distribution is not known, and consequently a set $\mathfrak{M}$ of probability measures (distributions) is constructed in some way and referred to as the *ambiguity set*. In that framework we can view (4.4) as the *definition* of the corresponding functional $\mathcal{R}$. It is straightforward to verify that the obtained functional $\mathcal{R}$, defined on an appropriate space of random variables, satisfies the axioms of coherent risk measures. So we have a certain duality relation between the risk averse and distributionally robust approaches to stochastic optimization. However, there is an essential difference in a way how the corresponding ambiguity set $\mathfrak{M}$ is constructed. In the risk averse approach the set $\mathfrak{M}$ consists of probability measures *absolutely continuous* with respect a specified (reference) measure $\mathbb{P}$. On the other hand, in the DRO approach such dominating measure (i.e., probability measure such that any measure in the set $\mathfrak{M}$ is absolutely continuous with respect that reference measure), may not be naturally defined. This happens, for

---

[2]By writing $\mathbb{E}_P[Z]$ we emphasize that the expectation is taken with respect to the probability measure (probability distribution) $P$.

instance, in popular settings where the ambiguity set is defined by moment constraints or is the set of probability measures with prescribed Wasserstein distance from the empirical distribution defined by the data.

In order to extend the risk averse and distributionally robust optimization to the dynamical setting of multistage stochastic optimization, we need to construct a conditional counterpart of the respective functional $\mathcal{R}$. In writing dynamic programming (2.4) and (3.7) we used conditional expectations in a somewhat informal manner. For the expectation operator there is a classical definition of conditional expectation. That is, let $P$ be a probability measure and $\mathcal{G}$ be a sigma subalgebra of $\mathcal{F}$. It is said that random variable, denoted $\mathbb{E}_{|\mathcal{G}}[Z]$, is the conditional expected value of random variable $Z$ given $\mathcal{G}$ if the following two properties hold: (i) $\mathbb{E}_{|\mathcal{G}}[Z] : \Omega \to \mathbb{R}$ is $\mathcal{G}$-measurable, (ii) $\int_A \mathbb{E}_{|\mathcal{G}}[Z](\omega)dP(\omega) = \int_A Z(\omega)dP(\omega)$ for any $A \in \mathcal{G}$. There are many versions of $\mathbb{E}_{|\mathcal{G}}[Z](\omega)$ which can differ from each other on sets of $P$-measure zero. The conditional expectation depends on measure $P$, therefore we sometimes write $\mathbb{E}_{P|\mathcal{G}}[Z]$ to emphasize this when various probability measures are considered.

It seems that it is natural to define the conditional counterpart of the distributionally robust functional, defined in (4.4), as $\sup_{P \in \mathfrak{M}} \mathbb{E}_{P|\mathcal{G}}[Z]$. However there are technical issues with a precise meaning of such definition since there are different versions of conditional expectations related to different probability measures $P \in \mathfrak{M}$. Even more importantly, there are conceptual problems with such definition, and in fact this is not how conditional risk measures are defined. A thorough exposition of this issue is beyond the scope of this monograph, the interested reader is referred to [53] for a careful discussion of this topic.

In the risk averse setting there is a natural concept of law invariance. For a random variable $Z$ its cumulative distribution function (cdf), with respect to the reference probability measure $\mathbb{P}$, is $F_Z(z) := \mathbb{P}(Z \leq z)$, $z \in \mathbb{R}$. It is said that random variables $Z$ and $Z'$ are *distributionally equivalent* if their cumulative distribution functions $F_Z(z)$ and $F_{Z'}(z)$ do coincide for all $z \in \mathbb{R}$. It is said that risk measure $\mathcal{R}$ is *law invariant* if $\mathcal{R}(Z) = \mathcal{R}(Z')$ for any distributionally equivalent $Z$ and $Z'$. For example the Average Value-at-Risk measure $\mathcal{R} = \mathsf{AV@R}_\alpha$ is law invariant.

Note that law invariance is defined with respect to the reference measure $\mathbb{P}$. Law invariant coherent risk measure $\mathcal{R}(Z)$ can be considered as a function of the cumulative distribution function $F_Z$, and consequently its conditional counterpart can be defined as the respective function of the conditional counterpart of $F_Z$; this definition can be made precise. On the other hand, in the DRO setting where there is no dominating probability measure and the concept of law invariance is not applicable, it is not completely clear how to define a conditional counterpart of functional $\mathcal{R}$ defined in (4.4), although an attempt was made in [53]. See Remark 5.7 below for a further discussion of law invariance.

In the risk neutral setting we were particularly interested in the stagewise independent case (Assumption 2.2). Under that assumption the dynamic programming equations simplify to (2.8)-(2.9) in the SP and to (3.11) in the SOC frameworks, respectively. Natural counterparts of these equations are obtained by replacing the expectation operator with a coherent risk measure. That is, in the risk averse SP framework the counterpart of (2.9) is

$$\mathcal{Q}_{t+1}(x_t) := \mathcal{R}_{t+1}[Q_{t+1}(x_t, \xi_{t+1})], \ t = 1, ..., T - 1, \qquad (4.5)$$

where $\mathcal{R}_{t+1}$ is a coherent risk measure applied to the random variable $Q_{t+1}(x_t, \xi_{t+1})$. For example,

$$\mathcal{R}_t(\cdot) := (1 - \lambda_t)\mathbb{E}[\,\cdot\,] + \lambda_t \mathsf{AV@R}_{\alpha_t}(\cdot), \ \lambda_t \in [0, 1], \qquad (4.6)$$

is a law invariant coherent risk measure representing a convex combination of the expectation and Average Value-at-Risk. Such choice of risk measure tries to reach a compromise between minimization of the cost on average and controlling the risk of upper deviations from the average (upper quantiles of the cost) at every stage of the process. In applications the parameters $\lambda_t$ and $\alpha_t$ often are constants, independent of the stage $t$, i.e., the risk measure $\mathcal{R}_t = \mathcal{R}$ is the same for every stage (cf., [71]).

**Remark 4.1.** The dynamic programming equations, with (4.5) replacing (2.9), correspond to the *nested* formulation of the risk averse multistage SP problem (cf., [62],[68, Section 6.5]). In the nested risk averse multistage optimization the risk is controlled at *every stage* of the process and the total nested value of the cost does not have practical interpretation. Nevertheless, when solving the corresponding risk averse optimization problem it would be useful to estimate its optimal value for the purpose of evaluating error of the computed solution. $\qquad\square$

In the DRO setting the counterpart of (2.9) can be written as

$$\mathcal{Q}_{t+1}(x_t) := \sup_{P \in \mathfrak{M}_{t+1}} \mathbb{E}_P[\mathcal{Q}_{t+1}(x_t, \xi_{t+1})], \qquad (4.7)$$

where $\mathfrak{M}_{t+1}$ is a set of probability distributions of vector $\xi_{t+1}$. Note that by convexity and monotonicity properties of coherent risk measures, convexity of the cost-to-go functions is preserved here. That is, convexity of $\mathcal{Q}_{t+1}(\cdot, \xi_{t+1})$ implies convexity of $\mathcal{Q}_{t+1}(\cdot)$ in the left hand side of (4.5) and (4.7). This in turn implies convexity of $\mathcal{Q}_t(\cdot, \xi_t)$ in the left hand side of (2.8), provided the function $f_t(\cdot, \xi_t)$ is convex and the set $\mathcal{X}_t$ is convex. Therefore we have here similar to the risk neutral case (compare with Proposition 2.1) the following convexity property.

**Proposition 4.1.** If the objective functions $f_t(x_t, \xi_t)$ are convex in $x_t$ and the sets $\mathcal{X}_t$ are convex for $t = 1, ..., T$, then the cost-to-go functions $\mathcal{Q}_{t+1}(x_t, \xi_{t+1})$ and $\mathcal{Q}_{t+1}(x_t)$ are convex in $x_t$. In particular such convexity of the cost-to-go functions follows for risk averse linear multistage stochastic programs.

Similar considerations apply to the SOC framework, with the risk averse counterpart of (3.11) written as

$$V_t(x_t) = \inf_{u_t \in \mathcal{U}_t(x_t)} \mathcal{R}_t \left[ c_t(x_t, u_t, \xi_t) + V_{t+1}(F_t(x_t, u_t, \xi_t)) \right], \qquad (4.8)$$

where $\mathcal{R}_t$, $t = 1, ..., T$, are coherent risk measures. Again these dynamic programming equations correspond to the *nested* formulation of the respective SOC problem. Similar to the risk neutral case, the value functions $V_t(x_t)$ are convex if Assumption 3.1 holds.

# 5

# Dynamic Cutting Plane Algorithms

The basic idea of cutting plane algorithms for solving convex problems is approximation of the (convex) objective function by cutting planes. That is, let $f : \mathbb{R}^n \to \mathbb{R} \cup \{+\infty\}$ be an extended real valued function. Its domain is $\text{dom}(f) := \{x \in \mathbb{R}^n : f(x) < +\infty\}$. We say that for some $\alpha \in \mathbb{R}$ and $\beta \in \mathbb{R}^n$, an affine function $\ell(x) = \alpha + \beta^\top x$ is a *cutting plane* of $f(x)$ if $f(x) \geq \ell(x)$ for all $x \in \mathbb{R}^n$. If moreover $f(\bar{x}) = \ell(\bar{x})$ for some $\bar{x} \in \text{dom}(f)$, then $\ell(x)$ is said to be a *supporting plane* of $f(x)$, at the point $\bar{x}$. When the function $f$ is convex, its supporting plane at a point $\bar{x} \in \text{dom}(f)$ is given by $\ell(x) = f(\bar{x}) + g^\top(x - \bar{x})$, where $g$ is a subgradient of $f$ at $\bar{x}$. The set of all subgradients, at a point $x \in \text{dom}(f)$, is called the subdifferential of $f$ at $x$ and denoted $\partial f(x)$. The subdifferential of a convex function at every interior point of its domain is nonempty, [61, Theorem 23.4].

Stochastic Dual Dynamic Programming (SDDP) is a cutting plane type algorithm for multistage linear SP problems introduced in Pereira and Pinto [47] and built on the nested decomposition algorithm of Birge [5]. Each iteration of the SDDP algorithm goes backward and forward through different stages in order to build piecewise linear approximations of the cost-to-go functions by their cutting planes.

**Remark 5.1.** Let us recall how to compute a subgradient of the optimal value function of a linear program. That is, let $Q(x)$ be the optimal value of linear program,

$$\min_{y \geq 0} c^\top y \ \text{ s.t. } \ Bx + Ay = b, \tag{5.1}$$

which is considered as a function[1] of vector $x$. The dual of (5.1) is the linear program

$$\max_{\lambda} \lambda^\top (b - Bx) \ \text{ s.t. } \ A^\top \lambda \leq c. \tag{5.2}$$

Function $Q(\cdot)$ is piecewise linear convex, and its subgradient at a point $x$ where $Q(x)$ is finite, is given by $-B^\top \bar{\lambda}$ with $\bar{\lambda}$ being an optimal solution of the dual problem (5.2) (e.g., [68, Section 2.1.1]). That is, the subgradient of the optimal value function of a linear program is computed by solving the dual of that linear program. A more general setting amendable to linear programming formulation is when the objective function of (5.1) is polyhedral (i.e., its epigraph is a polyhedron). In such cases the gradient of the respective optimal value function can be computed by solving the dual of the obtained linear program. This principle is applied in the SDDP algorithm to approximate the cost-to-go functions by their cutting planes in a dynamical setting. This motivates the "Dual Dynamic" part in the name of the algorithm.  □

## 5.1   SDDP Algorithm for SP Problems

Consider the multistage SP problem in (2.1)-(2.2). Suppose that Assumptions 2.1 and 2.2 hold, and hence the corresponding cost-to-go functions are defined by equations (2.8)-(2.9). If the probability distributions of random vectors $\xi_t$ are continuous, these distributions should be discretized in order to compute the respective expectations (see Remark 5.2 below). So we also make the following assumption.

**Assumption 5.1.** The probability distribution of $\xi_t$ has finite support $\{\xi_{t1}, ..., \xi_{tN}\}$ with respective probabilities[2] $p_{t1}, ..., p_{tN}$, $t = 2, ..., T$. De-

---

[1] If for some $x$ problem (5.1) does not have a feasible solution, then $Q(x) := +\infty$.

[2] For the sake of simplicity we assume that the number $N$ of realizations of $\xi_t$ is the same for every stage $t = 2, ..., T$. Recall that in the SP framework, $\xi_1$ is deterministic.

note $A_{tj} := A_t(\xi_{tj})$, $B_{tj} := B_t(\xi_{tj})$, $b_{tj} := b_t(\xi_{tj})$, and $Q_{tj}(x_{t-1}) := Q_t(x_{t-1}, \xi_{tj})$ the cost-to-go functions, defined in (2.8).

The corresponding expected value cost-to-go functions can be written as

$$\mathcal{Q}_t(x_{t-1}) = \sum_{j=1}^{N} p_{tj} Q_{tj}(x_{t-1}). \tag{5.3}$$

Suppose further that the objective functions

$$f_t(x_t, \xi_{tj}) := c_{tj}^\top x_t, \ t = 1, ..., T, \tag{5.4}$$

are linear in $x_t$, and the sets $\mathcal{X}_t := \mathbb{R}_+^{n_t}$, i.e., the constraint $x_t \in \mathcal{X}_t$, can be written as $x_t \geq 0$, $t = 1, ..., T$. In that case (2.1)-(2.2) becomes the standard linear multistage SP problem:

$$\min_{x_t \geq 0} \mathbb{E}\left[ \sum_{t=1}^{T} c_t^\top x_t \right] \text{ s.t. } B_t x_{t-1} + A_t x_t = b_t, \ t = 1, ..., T. \tag{5.5}$$

It should be noted that if the sets $\mathcal{X}_t$ are polyhedral and the functions $f_t(x_t, \xi_t)$ are polyhedral in $x_t$, then the problem is linear and can be formulated in the standard form by the well known techniques of linear programming.

**Remark 5.2.** From the modeling point of view, random vectors $\xi_t$ often are assumed to have continuous distributions. In such cases these continuous distributions have to be discretized in order to perform numerical calculations. One possible approach to such discretization is to use Monte Carlo sampling techniques. That is, a random sample $\xi_{t1}, ..., \xi_{tN}$ of size $N$ is generated from the probability distribution of random vector $\xi_t$, $t = 2, ..., T$. This approach is often referred to as the Sample Average Approximation (SAA) method. The constructed discretized problem can be considered in the above framework with equal probabilities $p_{tj} = 1/N$, $j = 1, ..., N$. Statistical properties of the SAA method in the multistage setting are discussed, e.g., in [68, Section 5.8]. It should be mentioned that the question of the sample size $N$, i.e., how many discretization points per stage to use, can be only addressed in the framework of the original model with continuous distributions. One of the advantages of the SAA method is that statistical tests can be applied to the results of computational procedures, e.g., whether there is a significant change in the SAA discretization with increase of the sample size $N$, see discussions in [11].                                     □

The SDDP algorithm for linear multistage SP problems, having finite number of scenarios, is described in detail in a number of publications (see, e.g., a recent survey paper [16] and references therein). For a later reference we briefly discuss it below. The SDDP algorithm consists of backward and forward steps. At each iteration of the algorithm, in the backward step the current cutting plane approximations $\underline{\mathcal{Q}}_t(\cdot)$ of the cost-to-go functions $\mathcal{Q}_t(\cdot)$ are updated by adding respective cuts in each stage (see Figure 5.1 for an illustration on the last two stages). That is, at the last stage the cost-to-go function $Q_{Tj}(x_{T-1})$, $j = 1, ..., N$, is given by the optimal solution of the linear program

$$\min_{x_T \geq 0} c_{Tj}^\top x_T \quad \text{s.t.} \quad B_{Tj} x_{T-1} + A_{Tj} x_T = b_{Tj}. \tag{5.6}$$

The subgradient of $Q_{Tj}(x_{T-1})$ at a chosen point[3] $\tilde{x}_{T-1}$ is given by $g_{Tj} = -B_{Tj}^\top \tilde{\lambda}_{Tj}$, where $\tilde{\lambda}_{Tj}$ is an optimal solution of the dual of problem (5.6) (see Remark 5.1 above). Consequently, the subgradient of $\mathcal{Q}_T(x_{T-1})$ at $\tilde{x}_{T-1}$ is computed as $\mathbf{g}_T = \sum_{j=1}^N p_{Tj} g_{Tj}$, and hence the cutting plane

$$\ell_T(x_{T-1}) = \mathcal{Q}_T(\tilde{x}_{T-1}) + \mathbf{g}_T^\top(x_{T-1} - \tilde{x}_{T-1}) \tag{5.7}$$

is added to the current family of cutting planes of $\mathcal{Q}_T(\cdot)$. Actually at the last stage the above cutting plane $\ell_T(\cdot)$ is the *supporting* plane of $\mathcal{Q}_T(\cdot)$ at $\tilde{x}_{T-1}$. This is because $\mathcal{Q}_T(\tilde{x}_{T-1})$ is explicitly computed by solving problem (5.6).

At one step back at stage $T-1$, the cost-to-go function $Q_{T-1,j}(x_{T-2})$, $j = 1, ..., N$, is given by the optimal solution of the program

$$\min_{x_{T-1} \geq 0} c_{T-1,j}^\top x_{T-1} + \mathcal{Q}_T(x_{T-1}) \quad \text{s.t.} \quad B_{T-1,j} x_{T-2} + A_{T-1j} x_{T-1} = b_{T-1,j}.$$
$$\tag{5.8}$$

The cost-to-go function $\mathcal{Q}_T(\cdot)$ is not known. Therefore it is replaced by the approximation $\underline{\mathcal{Q}}_T(\cdot)$ given by the maximum of the current family of cutting planes including the cutting plane added at stage $T$ as described

---

[3]Choice and meaning of the so-called trial points $\tilde{x}_t$ will be discussed later (in particular see Remark 5.4 below and Section 6).

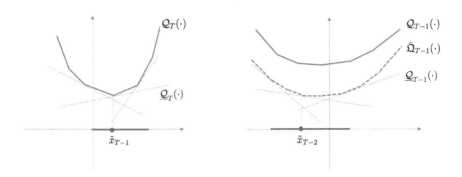

**Figure 5.1:** Backward step of SDDP for the last two stages.

above. Since $\underline{\mathcal{Q}}_T(\cdot)$ is the maximum of a finite number of affine functions, the obtained problems

$$\min_{x_{T-1}\geq 0} c_{T-1,j}^\top x_{T-1} + \underline{\mathcal{Q}}_T(x_{T-1}) \quad \text{s.t.} \quad B_{T-1,j} x_{T-2} + A_{T-1j} x_{T-1} = b_{T-1,j},$$
$$(5.9)$$

$j = 1, ..., N$, are linear. Let $\underline{\mathfrak{Q}}_{T-1,j}(x_{T-2})$ be the optimal value function of problem (5.9) and $\bar{\underline{\mathfrak{Q}}}_{T-1}(x_{T-2}) = \sum_{j=1}^N p_{T-1,j} \underline{\mathfrak{Q}}_{T-1,j}(x_{T-2})$ be the corresponding estimate of the expected value cost function. Note that since $\mathcal{Q}_T(\cdot) \geq \underline{\mathcal{Q}}_T(\cdot)$, we have that $Q_{T-1,j}(\cdot) \geq \underline{\mathfrak{Q}}_{T-1,j}(\cdot)$, $j = 1, ..., N$, and hence $\mathcal{Q}_{T-1}(\cdot) \geq \bar{\underline{\mathfrak{Q}}}_{T-1}(\cdot)$. Gradient $g_{T-1,j}$ of the optimal value function $\underline{\mathfrak{Q}}_{T-1,j}(x_{T-2})$ can be computed at a chosen point $\tilde{x}_{T-2}$ by solving the dual of problem (5.9) and hence computing its optimal solution. Then $g_{T-1} = \sum_{j=1}^N p_{T-1,j} g_{T-1,j}$ is the gradient of $\bar{\underline{\mathfrak{Q}}}_{T-1}(x_{T-2})$ at $\tilde{x}_{T-2}$. Consequently

$$\ell_{T-1}(x_{T-2}) = \underline{\mathcal{Q}}_{T-1}(\tilde{x}_{T-2}) + g_{T-1}^\top(x_{T-2} - \tilde{x}_{T-2}) \qquad (5.10)$$

is a cutting plane of $\mathcal{Q}_{T-1}(\cdot)$, and is added to the current collection of cutting planes of $\mathcal{Q}_{T-1}(\cdot)$. And so going backward in time, $\underline{\mathcal{Q}}_t(\cdot)$ are defined as the maximum over generated cutting planes until the corresponding approximation of the first stage problem is obtained:

$$\min_{x_1 \geq 0} c_1^\top x_1 + \underline{\mathcal{Q}}_2(x_1) \quad \text{s.t.} \quad A_1 x_1 = b_1. \qquad (5.11)$$

Let us note that by the construction, $\mathcal{Q}_t(\cdot) \geq \underline{\mathcal{Q}}_t(\cdot)$ for $t = 2, ..., T$. Therefore any cutting plane of $\underline{\mathcal{Q}}_t(\cdot)$ is also a cutting plane of $\mathcal{Q}_t(\cdot)$.

Also, it follows that the optimal value of problem (5.11) is less than or equal to the optimal value of (2.1)-(2.2). Therefore, the optimal value of (5.11) provides a (deterministic) lower bound for the optimal value of the considered multistage problem.

The computed estimates $\underline{\mathcal{Q}}_t(\cdot)$, $t = 2, ..., T$, of the cost-to-go functions, and optimal solution $\bar{x}_1$ of the first stage problem (5.11), define a feasible policy in a way similar to (2.10), obtained by replacing $\mathcal{Q}_{t+1}(\cdot)$ in (2.10) with $\underline{\mathcal{Q}}_{t+1}(\cdot)$. Since this policy is feasible, its value is greater than or equal to the optimal value of the considered problems (2.1)-(2.2). In the forward step this value is estimated by randomly generating realizations of the data process and evaluating the corresponding policy values. That is, consider a sample path (scenario) $\hat{\xi}_2, ..., \hat{\xi}_T$ of the data process. Let $\hat{B}_t = B_t(\hat{\xi}_t)$, $\hat{A}_t = A_t(\hat{\xi}_t)$, $\hat{b}_t = b_t(\hat{\xi}_t)$ and $\hat{c}_t = c_t(\hat{\xi}_t)$ be realizations of the random parameters corresponding to the generated sample path. Starting with first stage solution $\bar{x}_1$, next values $\bar{x}_t$ are computed iteratively going forward in time by computing an optimal solution of the optimization problem in (2.10) with $\mathcal{Q}_{t+1}(\cdot)$ replaced by $\underline{\mathcal{Q}}_{t+1}(\cdot)$. That is

$$\bar{x}_t \in \arg\min_{x_t \geq 0} \left\{ \hat{c}_t^\top x_t + \underline{\mathcal{Q}}_{t+1}(x_t) : \hat{B}_t \bar{x}_{t-1} + \hat{A}_t x_t = \hat{b}_t \right\}. \qquad (5.12)$$

The corresponding total sum

$$\hat{\vartheta} := \hat{c}_1^\top \bar{x}_1 + \sum_{t=2}^{T} \hat{c}_t^\top \bar{x}_t \qquad (5.13)$$

is a function of the generated sample and hence is random. Its expected value $\mathbb{E}[\hat{\vartheta}]$ is equal to the value of the constructed policy, and thus $\mathbb{E}[\hat{\vartheta}]$ is greater than or equal to the optimal value of (2.1)-(2.2).

The forward step of the algorithm proceeds to evaluate the value of the constructed policy by randomly generating the sample paths (scenarios). At every stage $t = 2, ..., T$, a point $\xi_t$ is generated at random according to the distribution of the corresponding random vector. That is, a point $\hat{\xi}_t$ is chosen with probability $p_{ti}$, $i = 1, ..., N$, from the set $\{\xi_{t1}, ..., \xi_{tN}\}$. This generates the corresponding sample path (scenario) $\hat{\xi}_2, ..., \hat{\xi}_T$ of the data process. Consequently the policy values for the generated sample path are computed, by using the rule

(5.12), and thus the total sum $\hat{\vartheta}$, defined in (5.13), is evaluated. The computed value $\hat{\vartheta}$ gives a point estimate of value of the constructed policy. This procedure can be repeated several times by randomly (and independently from each other) generating $M$ scenarios. Let $\hat{\vartheta}^1, ..., \hat{\vartheta}^M$ be values, calculated in accordance with (5.13), for each generated scenario. Their average $\bar{\vartheta} := M^{-1} \sum_{i=1}^{M} \hat{\vartheta}^i$ gives an unbiased estimate of the value of the considered policy. An upper edge $\bar{\vartheta} + z_\alpha S/\sqrt{M}$ of the corresponding confidence interval is viewed as a statistical estimate of an upper bound for the optimal value of the considered multistage problem. Here $S^2 = (M - 1)^{-1} \sum_{i=1}^{M} (\hat{\vartheta}^i - \bar{\vartheta})^2$ is an estimate of the variance $\mathrm{Var}(\hat{\vartheta})$, and $z_\alpha$ is the critical level, say $z_\alpha = 2$. The justification for this bound is that by the Central Limit Theorem (CLT) for large $M$, the average $\bar{\vartheta}$ has approximately normal distribution and $z_\alpha$ corresponds to mass $100(1-\alpha)\%$ of the standard normal distribution. For $M$ not too large it is a common practice to use critical values from $t$-distribution with $M - 1$ degrees of freedom, rather than normal. In any case this is just a heuristic justified by the CLT. Nevertheless it is useful and typically gives a reasonable estimate of the upper bound.

**Remark 5.3.** The constructed approximations $\mathcal{Q}_t(\cdot)$ of the expected value cost-to-go functions also define a feasible policy for the original problem with continuous distributions of the data process vectors $\xi_t$. Its value can be estimated in the same way by generating a sample path $\hat{\xi}_2, ..., \hat{\xi}_T$ from the *original* continuous distribution and computing the corresponding policy values in the forward step in accordance with (5.12). Consequently the policy value is estimated by averaging the computed values for a number of randomly generated sample paths. This can be used for the construction of statistical upper bound for the original problem. Expectation of the optimal value of the SAA problem, based on randomly generated sample, is less than or equal to the optimal value of the original problem. Therefore the lower bound generated by the SDDP algorithm gives a point estimate of a lower bound of the original problem. We can refer to [11] for various related statistical testing procedures. □

The difference between the (statistical) upper bound and (determin- istic) lower bound provides an upper bound for the optimality gap of the constructed policy. It can be used as a stopping rule by stopping the iteration procedure when this estimated gap is smaller than a specified precision level. The bound is quite conservative and for larger problems may not be reducible to the specified accuracy level in a reasonable computational time. A more practical heuristic approach is to stop the iterations after the lower bound stabilizes and new iterations, with additional cutting planes, do not significantly improve (increase) the lower bound.

**Remark 5.4.** The above procedure depends on a choice of the points $\tilde{x}_t$ referred to as the *trial points* (see Figure 5.1 and the discussion afterwards). The forward step suggests to use the computed values $\bar{x}_t$ (see (5.12)) as trial points in the next iteration of the backward step of the algorithm. The rationale for this can be explained as follows. In general, in order to approximate the cost-to-go functions *uniformly* with a given accuracy in a specified region, the number of cutting planes grows exponentially with increase of the dimension of decision variables $x_t$. In many applications the regions where the cost-to-go functions should be approximated more accurately are considerably smaller than the regions of possible variability of decision variables. Utilizing the most recently generated cutting plane models, the forward step is supposed to generate new trial points where likely the cost-to-go functions should be approximated more accurately. In Section 6, we will investigate further how the rate of convergence of SDDP type algorithms is impacted by the selection of trial points generated in the forward step.          □

**Remark 5.5.** The trial points are supposed to be chosen at stages $t = 2, ..., T - 1$. This suggests that the above procedure is relevant for number of stages $T \geq 3$. For two stage problems, i.e., when $T = 2$, there is no sampling involved and this becomes the classical Kelley's cutting plane algorithm [26]. It is well known in the deterministic convex optimization that Kelley's algorithm can behave poorly with increase in the number of decision variables, and much better cutting plane algorithms can be designed by using regularization or trust region/level set strategies [28], [36]. However, it is not clear how such methods can be extended to the dynamic multistage settings (although such attempts were made). The reason is that in the multistage problems the optimization is performed over policies which are functions of the data process. It is not known a priori where realizations of considered policies may happen in future stages and an attempt to restrict regions of search by some form of regularization could be not helpful. □

Proofs of convergence of the SDDP type algorithms were studied in several publications (e.g., [18] and references therein). These proofs are based on the argument that the forward step of the algorithm will eventually almost surely go through every possible scenario many times, and hence the cost-to-go functions will be approximated with any accuracy at relevant regions. Since the number of scenarios grows exponentially with respect to the number of stages $T$, SDDP may converge very slowly as $T$ increases. Numerical experiments indicate that, in a certain sense, the numerical complexity of SDDP method grows almost linearly with increase of the number of stages while the dimension of decision vectors is constant. The rate of convergence of SDDP and its certain variants was investigated in a recent paper [31] (see Section 6).

### 5.1.1 Interstage dependence

In many applications the assumption of stagewise independence (Assumption 2.2) is unrealistic. Suppose now that the data process $\xi_t \in \mathbb{R}^d$ is Markovian (see Remark 2.3). In that case the cost-to-go functions are defined by (2.11)-(2.12). There are different ways how such Markovian structure can be modeled (e.g., [7], [38], [39], [51], [52], [72]). One classical approach is time series analysis. The data process is modeled as an, say first order, autoregressive process, i.e.,

$$\xi_t = \mu + \Phi\xi_{t-1} + \varepsilon_t, \tag{5.14}$$

where $\Phi \in \mathbb{R}^{d \times d}$ and $\mu \in \mathbb{R}^d$ are parameters of the process and $\varepsilon_t$ is a sequence of random error vectors. It is assumed that the error process $\varepsilon_t$ is stagewise independent (in fact it is usually assumed that $\varepsilon_t$ are i.i.d.). Suppose further that only the right hand side variables $b_t$ in the considered problem (2.11)-(2.12) are random and set $\xi_t := b_t$. Then we can write the cost-to-go functions in terms of $x_t$ and $\xi_t$, treating $\varepsilon_t$ as the corresponding random process:

$$Q_t(x_{t-1}, \xi_{t-1}, \varepsilon_t) = \inf_{x_t \geq 0, \xi_t} \left\{ c_t^\top x_t + \mathcal{Q}_{t+1}(x_t, \xi_t) : \begin{array}{l} B_t x_{t-1} + A_t x_t = \xi_t, \\ \xi_t - \mu - \Phi \xi_{t-1} = \varepsilon_t \end{array} \right\}, \tag{5.15}$$

with

$$\mathcal{Q}_{t+1}(x_t, \xi_t) = \mathbb{E}[Q_{t+1}(x_t, \xi_t, \varepsilon_{t+1})]. \tag{5.16}$$

Here $(x_t, \xi_t)$ are treated as decision variables and $\varepsilon_t$ is viewed as the corresponding random process with the expectation in (5.16) taken with respect to $\varepsilon_{t+1}$. (5.15)-(5.16) can be viewed in the framework of dynamic programming equations (2.8)-(2.9), and hence the standard SDDP algorithm can be directly applied.

There are two drawbacks in the above approach. First, the number of decision variables is artificially increased with the cutting planes in the corresponding SDDP algorithm computed jointly in $x_t$ and $\xi_t$. This increases the computational complexity of the numerical procedure. Second, in order to have a *linear* problem, only the right hand side parameters are allowed to be random. An alternative approach is to approximate the Markovian process $\xi_t$ by Markov chain, which we are going to discuss next.

Suppose that the distribution of random vector $\xi_t$ is supported on set $\Xi_t \subset \mathbb{R}^{d_t}$, $t = 2, ..., T$. Consider the partition[4] $\Xi_{ti}$, $i = 1, ..., N$, of $\Xi_t$ by Voronoi cells, i.e.,

$$\Xi_{ti} := \{\xi_t \in \Xi_t : \|\xi_t - \eta_{ti}\| \leq \|\xi_t - \eta_{tj}\| \text{ for all } j \neq i\},$$

where $\eta_{ti} \in \Xi_t$ are given points. That is, $\Xi_{ti}$ is the set of points of $\Xi_t$ closest to $\eta_{ti}$. There are various ways how the center points $\eta_{ti}$ can be chosen. For example, $\eta_{ti}$ could be determined by minimizing the squared distances:

$$\min_{\eta_{t1}, ..., \eta_{tN}} \int \min_{i=1, ..., N} \|\xi_t - \eta_{ti}\|^2 dP(\xi_t). \tag{5.17}$$

The above optimization problem is nonconvex. Nevertheless it can be reasonably solved by various methods (cf., [7], [38]). Consider the conditional probabilities

$$p_{tij} := P(\xi_{t+1} \in \Xi_{t+1,j} | \xi_t \in \Xi_{ti}).$$

---

[4]For the sake of simplicity we assume that the number $N$ of cells is the same at every stage.

Note that different Voronoi cells can have common boundary points. We assume that probabilities of such events are zero so that the probabilities $p_{tij}$ are well defined. This leads to an approximation of the data process by a (possibly nonhomogeneous) Markov chain with transition probabilities $p_{tij}$ of moving from point $\eta_{ti}$ to point $\eta_{t+1,j}$ at the next stage.

Consider $c_{ti} := c_t(\eta_{ti})$, $A_{ti} := A_t(\eta_{ti})$, $B_{ti} := B_t(\eta_{ti})$ and $b_{ti} := b_t(\eta_{ti})$, $i = 1, ..., N$. Recall dynamic programming equations (2.11)-(2.12) for a Markovian data process. For the constructed Markov chain these dynamic programming equations can be written as

$$Q_t(x_{t-1}, \eta_{ti}) = \inf_{x_t \geq 0} \left\{ c_{ti}^\top x_t + \mathcal{Q}_{t+1}(x_t, \eta_{ti}) : B_{ti}x_{t-1} + A_{ti}x_t = b_{ti} \right\}, \; i = 1, ..., N,$$

(5.18)

where

$$\mathcal{Q}_{t+1}(x_t, \eta_{ti}) = \sum_{j=1}^{N} p_{tij} Q_{t+1}(x_t, \eta_{t+1,j}).$$ (5.19)

Note that the cost-to-go functions $Q_t(x_{t-1}, \eta_t)$ are convex in $x_{t-1}$, while the argument $\eta_t$ can take only the values $\eta_{t1}, ..., \eta_{tN}$.

It is possible to apply the SDDP method separately for every $\eta_t = \eta_{ti}$, $i = 1, ..., N$ (cf., [38]). That is, in the backward step of the algorithm the cost-to-go functions $\mathcal{Q}_{t+1,i}(x_t) := \mathcal{Q}_{t+1}(x_t, \eta_{ti})$ are approximated by their cutting planes (in $x_t$) separately for each $\eta_{ti}$, $i = 1, ..., N$. This defines a policy for every realization (sample path) $\eta_1, ...,$ of the constructed Markov chain. This can be used in the forward step of the algorithm.

It could happen that a sample path $\xi_1, ...,$ of the original data process is different from every sample path $\eta_1, ...,$ of the constructed Markov chain. In order to extend the constructed policy to a policy for the original data process, one approach is to define the policy values for a sample path of the original data process by taking the corresponding policy values from the sample path of the Markov chain closest in some sense to the considered sample path of the original data process.

The above approach is based on discretization (approximation) of the Markov process by a Markov chain. Compared with the approach based on the autoregressive modeling of the data process, the approach of Markov chain discretization is not restricted to the case where only the right hand side parameters are random. Also it uses cutting plane approximations only with respect to variables $x_t$, which could be computationally advantageous. On the other hand, its computational complexity and required computer memory grows as the number of discretization points per stage increases. Also unlike the SAA method, there is no available inference describing statistical properties of such discretization. Intuitively, such discretization could become problematic with increase of the dimension of the random vectors of the data process.

## Duality of multistage linear stochastic programs

The linear multistage stochastic program (5.5), with a finite number of scenarios, can be viewed as a large scale linear program, referred to as the deterministic equivalent. By the standard theory of linear programming it has a (Lagrangian) dual. Suppose that the primal problem (5.5) has finite optimal value. Then the optimal values of the primal and its dual are equal to each other and both problems have optimal solutions. Dualization of the feasibility constraints leads to the following (Lagrangian) dual of problem (5.5) (cf., [68, Section 3.2.3])

$$
\begin{aligned}
\max_{\pi} \quad & \mathbb{E}\big[\textstyle\sum_{t=1}^{T} b_t^\top \pi_t\big] \\
\text{s.t.} \quad & A_T^\top \pi_T \leq c_T, \\
& A_{t-1}^\top \pi_{t-1} + \mathbb{E}_{|\xi_{[t-1]}}\big[B_t^\top \pi_t\big] \leq c_{t-1},\ t = 2, ..., T.
\end{aligned}
\tag{5.20}
$$

The optimization in (5.20) is over policies[5] $\pi_t = \pi_t(\xi_{[t]})$, $t = 1, ..., T$.

It is possible to write dynamic programming equations for the dual problem. The idea of applying the SDDP algorithm to the dual problem was introduced in [35]. In what follows we use the approach of [21] based on (5.20). As before we make Assumptions 2.1, 2.2 and 5.1. At the last stage $t = T$, given $\pi_{T-1}$ and $\xi_{[T-1]}$, we need to solve the following problem with respect to $\pi_T$:

---

[5] Here policies $\pi$ and value functions $V_t$ should not be confused with the policies and value functions of the SOC framework.

$$\max_{\pi_{T1},\ldots,\pi_{TN}} \sum_{j=1}^{N} p_{Tj} b_{Tj}^{\top} \pi_{Tj}$$

$$\text{s.t.} \quad A_{Tj}^{\top} \pi_{Tj} \leq c_{Tj}, \; j = 1, \ldots, N, \tag{5.21}$$

$$A_{T-1}^{\top} \pi_{T-1} + \sum_{j=1}^{N} p_{Tj} B_{Tj}^{\top} \pi_{Tj} \leq c_{T-1}.$$

Note that $\sum_{j=1}^{N} p_{Tj} b_{Tj}^{\top} \pi_{Tj} = \mathbb{E}[b_T^{\top} \pi_T]$ and $\sum_{j=1}^{N} p_{Tj} B_{Tj}^{\top} \pi_{Tj} = \mathbb{E}\left[ B_T^{\top} \pi_T \right]$. Since $\xi_T$ is independent of $\xi_{[T-1]}$, the expectation in (5.21) is unconditional with respect to the distribution of $\xi_T$.

The optimal value $V_T(\pi_{T-1}, \xi_{T-1})$ and an optimal solution $(\bar{\pi}_{T1}, \ldots, \bar{\pi}_{TN})$ of (5.21) are functions of vectors $\pi_{T-1}$ and $c_{T-1} = c_{T-1}(\xi_{T-1})$ and matrix $A_{T-1} = A_{T-1}(\xi_{T-1})$. And so on going backward in time, using the stagewise independence assumption, we can write the respective dynamic programming equations for $t = T - 1, \ldots, 2$, as

$$\max_{\pi_{t1},\ldots,\pi_{tN}} \sum_{j=1}^{N} p_{tj} \left[ b_{tj}^{\top} \pi_{tj} + V_{t+1}(\pi_{tj}, \xi_{tj}) \right]$$

$$\text{s.t.} \quad A_{t-1}^{\top} \pi_{t-1} + \sum_{j=1}^{N} p_{tj} B_{tj}^{\top} \pi_{tj} \leq c_{t-1}, \tag{5.22}$$

with $V_t(\pi_{t-1}, \xi_{t-1})$ being the optimal value of (5.22). Finally at the first stage the following problem should be solved

$$\max_{\pi_1} b_1^{\top} \pi_1 + V_2(\pi_1, \xi_1). \tag{5.23}$$

Let us make the following observations about the dual problem. Unlike the primal problem, the optimization (maximization) (5.21) and (5.22) do not decompose into separate problems with respect to each $\pi_{tj}$ and should be solved as one linear program with respect to $(\pi_{t1}, \ldots, \pi_{tN})$. If $A_t$ and $c_t$, $t = 2, \ldots, T$, are deterministic, then $V_t(\pi_{t-1})$ is only a function of $\pi_{t-1}$. The value (cost-to-go) function $V_t(\pi_{t-1}, \xi_{t-1})$ is a concave function of $\pi_{t-1}$. This allows to apply an SDDP type algorithm to the dual problem (see [21] for details). For any feasible policy of the dual problem, the value of the first stage problem (5.23) gives an upper bound for the optimal value of the primal problem (5.5). Since the dual is a maximization problem, the SDDP algorithm applied to the dual problem produces a deterministic upper bound for the optimal value of (5.5).

It is interesting to note that it could happen that the dual problem does not have relatively complete recourse even if the primal problem has it. It is said that the problem (primal or dual) has *relatively complete recourse*, if at every stage $t = 2, ..., T$, for any generated solutions by the forward process at the previous stages, the respective dynamic program has a feasible solution at the stage $t$ for every realization of the random data. Without relatively complete recourse it could happen at a certain stage that for some realizations of the data process, the corresponding optimization problem is infeasible and the forward process cannot continue. Note that the relatively complete recourse is related to the policy construction in the forward dynamic process and may not hold even if the problem has finite optimal value. It is still possible to proceed with construction of the dual upper bound by introducing a penalty term to the dual problem (cf., [21]).

## 5.2   Cutting Plane Algorithm for SOC Problems

Consider the SOC problems (3.1)-(3.3). Suppose that Assumptions 2.1 and 2.2 hold, and hence (3.11) for the value function follows. Suppose further that Assumption 3.1 is satisfied, and hence value functions $V_t(\cdot)$ are convex. Then a cutting plane algorithm of the SDDP type can be applied in a way similar to the SP setting. There are, however, some subtle differences which we are going to discuss next.

In the backward step of the algorithm an approximation $\underline{V}_t(\cdot)$ of the value function $V_t(\cdot)$ is constructed by maximum of a family of cutting planes, similar to the SP setting, going backward in time. Let $\tilde{x}_t$, $t = 2, ..., T+1$, be trial points, possibly generated by the forward step of the algorithm at the previous iteration. At the backward step, starting at time $t = T+1$ the cutting (supporting) plane $\ell_{T+1}(x_{T+1}) = c_{T+1}(\tilde{x}_{T+1}) + \mathbf{g}_{T+1}^\top(x_{T+1} - \tilde{x}_{T+1})$, where $\mathbf{g}_{T+1}$ is a subgradient of $c_{T+1}(\cdot)$ at $\tilde{x}_{T+1}$, is added to the current family of cutting planes of $V_{T+1}(\cdot)$. Going backward in time, at stage $t$ we need to compute a subgradient of the approximate value function

$$\inf_{u_t \in \mathcal{U}_t(x_t)} \sum_{j=1}^{N} p_{tj} \left[ c_{tj}(x_t, u_t) + \underline{V}_{t+1}(A_{tj}x_t + B_{tj}u_t + b_{tj}) \right] \qquad (5.24)$$

at the trial point $x_t = \tilde{x}_t$, where $\underline{V}_{t+1}(\cdot) = \max_{1 \le i \le M} \ell_{t+1,i}(\cdot)$ is the current approximation of $V_{t+1}(\cdot)$ by cutting planes $\ell_{t+1,i}(\cdot)$. The minimization problem (5.24) is linear if functions $c_{tj}(x_t, u_t)$ are linear (polyhedral) in $u_t$, and $\mathcal{U}_t(x_t)$ is of the form (3.10) with the corresponding mapping $g_t(x_t, u_t)$ being affine in $u_t$.

Let us first consider the case where $\mathcal{U}_t(x_t) \equiv \mathcal{U}_t$ does not depend on $x_t$. Then the required subgradient[6] is given by (cf., [59, Theorem 23.8])

$$g_t = \sum_{j=1}^{N} p_{tj} \left[ \nabla c_{tj}(\tilde{x}_t, \tilde{u}_t) + A_{tj}^\top \nabla \underline{V}_{t+1}(A_{tj}\tilde{x}_t + B_{tj}\tilde{u}_t + b_{tj}) \right], \quad (5.25)$$

where $\tilde{u}_t \in \mathcal{U}_t$ is a minimizer of (5.24) for $x_t = \tilde{x}_t$, and $\nabla c_{tj}(\tilde{x}_t, \tilde{u}_t)$ is a subgradient of $c_{tj}(\cdot, \tilde{u}_t)$ at $\tilde{x}_t$. We also need to compute a subgradient of $\nabla \underline{V}_{t+1}(\cdot)$ at the points $A_{tj}\tilde{x}_t + B_{tj}\tilde{u}_t + b_{tj}$, $j = 1, ..., N$. The subgradient $\nabla \underline{V}_{t+1}(\cdot)$ at a point $x_{t+1}$ is given by the gradient $\nabla \ell_{t+1,\nu}(x_{t+1})$ of its cutting plane with $\nu \in \{1, ..., M\}$ such that $\ell_{t+1,\nu}(x_{t+1}) = \underline{V}_{t+1}(x_{t+1})$, i.e., $\ell_{t+1,\nu}$ is the supporting plane of $\underline{V}_{t+1}(\cdot)$ at $x_{t+1}$. Of course, the gradient of affine function $\ell(x) = \alpha + \beta^\top x$ is $\nabla \ell(x) = \beta$ for any $x$. The cutting plane is then $\underline{V}_t(\tilde{x}_t) + g_t^\top(x_t - \tilde{x}_t)$, where

$$\underline{V}_t(\tilde{x}_t) = \sum_{j=1}^{N} p_{tj} \left[ c_{tj}(\tilde{x}_t, \tilde{u}_t) + \underline{V}_{t+1}(A_{tj}\tilde{x}_t + B_{tj}\tilde{u}_t + b_{tj}) \right].$$

It should be noted that in the present case the required subgradient, and hence the corresponding cutting plane, can be computed without solving the dual problem. When $\mathcal{U}_t(x_t)$ depends on $x_t$, calculation of a subgradient of function (5.24) could be more involved; nevertheless in some cases it is still possible to proceed. Suppose that $\mathcal{U}_t(x_t)$ is of the form (3.10) with the corresponding mapping $g_t(\cdot, \cdot)$ having convex components. That is, minimization problem (5.24) can be written as

$$\min_{u_t} \sum_{j=1}^{N} p_{tj} \left[ c_{tj}(x_t, u_t) + \underline{V}_{t+1}(A_{tj}x_t + B_{tj}u_t + b_{tj}) \right] \quad \text{s.t. } g_t(x_t, u_t) \le 0.$$
$$(5.26)$$

Then the required gradient is given by the gradient of the Lagrangian $L_t(x_t, \bar{u}_t, \bar{\lambda}_t)$ of problem (5.26) with $\bar{u}_t$ being the respective optimal

---

[6]With some abuse of the notation we denote by $\nabla f(x)$ the subgradient of convex function $f(x)$, and the respective gradient when $f(\cdot)$ is differentiable at $x$.

solution and $\bar{\lambda}_t$ being the corresponding Lagrange multipliers vector, provided the optimal solution and Lagrange multipliers are unique (e.g., [8, Theorem 4.24]).

The forward step of the algorithm is performed similarly to the SP setting. Starting with initial value $x_1 = \bar{x}_1$ and generated sample path $\hat{\xi}_t$, $t = 1, ...,$ the policy $\bar{u}_t = \pi_t(x_t)$ is computed iteratively going forward in time (compare with (3.12)). That is, at stage $t = 1, ...,$ given value $\bar{x}_t$ of the state vector, the control value is computed as

$$\bar{u}_t \in \arg\min_{u_t \in \mathcal{U}_t(\bar{x}_t)} \sum_{j=1}^{N} p_{tj} \left[ c_{tj}(\bar{x}_t, u_t) + \underline{V}_{t+1}(A_{tj}\bar{x}_t + B_{tj}u_t + b_{tj}) \right]. \quad (5.27)$$

Consequently the next stage state value is set to

$$\bar{x}_{t+1} = A_t(\hat{\xi}_t)\bar{x}_t + B_t(\hat{\xi}_t)\bar{u}_t + b_t(\hat{\xi}_t), \quad (5.28)$$

and so on. Recall that in the SOC framework, $\xi_1$ is also random. The computed values $\bar{x}_t$ of the state variables can be used as trial points in the next iteration of the backward step of the algorithm.

**Remark 5.6** (*Q*-factor approach). Consider the dynamic programming equations (3.11) with affine state equations of the form (3.4), and define[7]

$$Q_t(x_t, u_t) := \mathbb{E}\left[ c_t(x_t, u_t, \xi_t) + V_{t+1}(A_t x_t + B_t u_t + b_t) \right]. \quad (5.29)$$

We have that

$$V_t(x_t) = \inf_{u_t \in \mathcal{U}_t(x_t)} Q_t(x_t, u_t). \quad (5.30)$$

The dynamic programming equation (3.11) can be written in terms of $Q_t(x_t, u_t)$ as

$$Q_t(x_t, u_t) = \mathbb{E}\left[ c_t(x_t, u_t, \xi_t) + \inf_{u_{t+1} \in \mathcal{U}_{t+1}(A_t x_t + B_t u_t + b_t)} Q_{t+1}\left(A_t x_t + B_t u_t + b_t, u_{t+1}\right) \right]. \quad (5.31)$$

The optimal policy $\bar{u}_t = \pi_t(x_t)$, defined by (3.12), can be written in terms of the *Q*-functions as

$$\bar{u}_t \in \arg\min_{u_t \in \mathcal{U}_t(x_t)} Q_t(x_t, u_t). \quad (5.32)$$

---

[7]The *Q*-functions here should not be confused with the cost-to-go functions of the SP framework.

Under Assumption 3.1, functions $Q_t(x_t, u_t)$ are convex jointly in $x_t$ and $u_t$. Consequently the cutting plane algorithm can be applied directly to the dynamic programming equation (5.31). That is, in the backward step current lower approximations $\underline{Q}_t(x_t, u_t)$ of the respective functions $Q_t(x_t, u_t)$ are updated by adding cutting planes $\ell_t(x_t, u_t)$ computed at trial points $(\tilde{x}_t, \tilde{u}_t)$ generated in the forward step of the algorithm. The constructed approximations $\underline{Q}_t(\cdot, \cdot)$ are used in the forward step of the algorithm for generating a point estimate of the expected value of the corresponding policy for a randomly generated sample path. □

## 5.3 SDDP Algorithms for Risk Averse Problems

Cutting plane algorithms of SDDP type can be extended to risk averse problems (cf., [52], [65]). Let us start with the SP framework.

### Stochastic programming framework

We follow the assumptions and notation of Section 5.1 and assume that (2.9) for the cost-to-go functions is replaced by its risk averse counterpart (4.5). The backward step of the algorithm is performed similar to the risk neutral case. We only need to adjust the calculation of subgradients of the cutting plane approximations $\underline{\mathcal{Q}}_{t+1}(\cdot)$ of the cost-to-go functions $\mathcal{Q}_{t+1}(\cdot)$.

In numerical algorithms, we deal with discrete distributions having a finite support. So we assume in this section that $\Omega = \{\omega_1, ..., \omega_N\}$ is a finite space equipped with sigma algebra $\mathcal{F}$ of all subsets of $\Omega$ and reference probability measure $\mathbb{P}$ with the respective *nonzero* probabilities $p_i = \mathbb{P}(\{\omega_i\})$, $i = 1, ..., N$. For a random variable (a function) $Z : \Omega \to \mathbb{R}$, we denote $Z_i := Z(\omega_i)$, $i = 1, ..., N$.

**Remark 5.7.** Recall the definition of law invariant risk measures from Section 4. Law invariant coherent risk measures have naturally defined conditional counterparts used in the construction of the corresponding nested risk measures (see (5.54)-(5.55) below). Suppose for the moment that the probabilities $p_i$ are such that for $\mathcal{I}, \mathcal{J} \subset \{1, ..., N\}$ the equality $\sum_{i \in \mathcal{I}} p_i = \sum_{j \in \mathcal{J}} p_j$ holds only if $\mathcal{I} = \mathcal{J}$. Then two variables $Z, Z' : \Omega \to \mathbb{R}$ are distributionally equivalent iff they do coincide. Therefore in that case any risk measure defined on the space of variables $Z : \Omega \to \mathbb{R}$ is law invariant. In the theory of law invariant coherent risk measures it is usually assumed that the reference probability space $(\Omega, \mathcal{F}, \mathbb{P})$ is atomless; of course an atomless space cannot be finite. In the case of finite space $\Omega$ it makes sense to consider law invariant risk measures when all probabilities $p_i$ are equal to each other, i.e., $p_i = 1/N$, $i = 1, ..., N$. In that case $Z, Z' : \Omega \to \mathbb{R}$ are distributionally equivalent iff there exists a permutation $\pi : \Omega \to \Omega$ such that $Z'_i = Z_{\pi(i)}$, $i = 1, ..., N$. Anyway we will not restrict the following discussion to the case of all probabilities $p_i$ being equal to each other.                                                $\square$

In the considered setting, both the expectation $\mathbb{E}_{\mathbb{P}}[Z] = \sum_{i=1}^{N} p_i Z_i$ and the dual representation (4.1) of coherent risk measure for a random variable $Z$ can be written as

$$\mathcal{R}(Z) = \sup_{\zeta \in \mathfrak{A}} \textstyle\sum_{i=1}^{N} p_i \zeta_i Z_i, \tag{5.33}$$

where $\mathfrak{A}$ is a convex closed subset of $\mathbb{R}^N$ consisting of vectors $\zeta \geq 0$ such that $\sum_{i=1}^{N} p_i \zeta_i = 1$. Since it is assumed that $\mathcal{R}(Z)$ is real valued, the set $\mathfrak{A}$ is bounded and hence is compact.

The subdifferential of $\mathcal{R}(\cdot)$ at $Z$ is given by

$$\partial \mathcal{R}(Z) = \arg\max_{\zeta \in \mathfrak{A}} \textstyle\sum_{i=1}^{N} p_i \zeta_i Z_i. \tag{5.34}$$

Specific formulas for the subgradients, in various examples of coherent risk measures, are listed in [68, Section 6.3.2]. These formulas can be applied for computation of cutting planes in the backward step of the algorithm. Several such examples were discussed and implemented in an SDDP type algorithm in [71].

As discussed in Section 5.1, the forward step of an SDDP type algorithm for the risk neutral setting has two functions; namely, construction of statistical upper bounds and generation of trial points. The constructed approximations $\underline{\mathcal{Q}}_{t+1}(x_t)$ of the cost-to-go functions define a feasible policy in the same way as in the risk neutral case by using formula (5.12). Therefore the forward step can be used for generation of trial points similar to the risk neutral case. However, because of the nonlinearity of the risk measures, it does not produce an unbiased estimate of the risk averse value of the constructed policy and cannot be applied in a straightforward way for computing statistical upper bounds similar to the risk neutral case. At the moment there are no known efficient numerical algorithms for computing statistical upper bounds for risk averse problems in the SP framework. Interestingly, as we will argue below, in the SOC setting it is possible to construct reasonably efficient statistical upper bounds for a certain class of risk measures. It is shown in [9] how to derive deterministic upper bounds for linear risk averse problems by applying an SDDP type algorithm to a dual formulation.

### 5.3.1 Stochastic optimal control framework

As before we suppose that Assumptions 2.1 and 2.2 hold, and thus value functions $V_t(x_t)$, associated with coherent risk measure $\mathcal{R}$, are determined by the dynamic programming equations (4.8). Furthermore, we suppose that Assumption 3.1 is satisfied and hence value functions $V_t(x_t)$ are convex.

We consider in this section the following class of coherent risk measures. This class is quite broad and includes many risk measures used in practice (the presentation below is based on [22]). We assume that the considered risk measures can be represented as

$$\mathcal{R}_t(Z) := \inf_{\theta \in \Theta} \mathbb{E}_{\mathbb{P}_t}[\Psi(Z, \theta)], \quad Z \in \mathcal{Z}, \tag{5.35}$$

where $\Theta$ is a nonempty convex closed subset of a finite dimensional vector space, $\Psi : \mathbb{R} \times \Theta \to \mathbb{R}$ is a real valued function, and $\mathcal{Z}$ is a linear space of measurable functions. For the sake of simplicity, we assume that the function $\Psi(z, \theta)$ is the same at every stage, while the reference probability distribution $\mathbb{P}_t$ of $\xi_t$ can be different at different stages. We make the following assumptions about function $\Psi(z, \theta)$.

(i) For every $Z \in \mathcal{Z}$, the expectation in the right hand side of (5.35) is well defined and the infimum is finite valued.

(ii) Function $\Psi(z, \theta)$ is convex in $(z, \theta) \in \mathbb{R} \times \Theta$.

(iii) $\Psi(\cdot, \theta)$ is monotonically nondecreasing, i.e., if $z_1 \leq z_2$, then $\Psi(z_1, \theta) \leq \Psi(z_2, \theta)$ for all $\theta \in \Theta$.

Under these assumptions, the functional defined in (5.35) satisfies the axioms of convexity and monotonicity. Convex combination of expectation and Average Value-at-Risk, defined in (4.6), is of the form (5.35) with $\Theta := \mathbb{R}$, $\mathcal{Z} := L_1(\Omega, \mathcal{F}, \mathbb{P})$ and

$$\Psi(z, \theta) := (1 - \lambda_t)z + \lambda_t \left(\theta + (1 - \alpha_t)^{-1}[z - \theta]_+\right). \tag{5.36}$$

For risk measures of the form (5.35), dynamic programming equations (4.8) can be written as

$$V_t(x_t) = \underbrace{\inf_{u_t \in \mathcal{U}_t(x_t), \, \theta \in \Theta} \sum_{j=1}^{N} p_{tj} \Psi\big(c_{tj}(x_t, u_t) + V_{t+1}(A_{tj}x_t + B_{tj}u_t + b_{tj}), \theta\big)}_{\mathbb{E}_{\mathbb{P}_t}[\Psi(c_t(x_t, u_t, \xi_t) + V_{t+1}(A_t x_t + B_t u_t + b_t), \theta)]}.$$

$$\tag{5.37}$$

Note that the minimization in the right hand side of (5.37) is performed jointly in controls $u_t$ and parameter vector $\theta$.

The backward step of the algorithm can be performed similar to the risk neutral case. In case $\mathcal{U}_t(x_t) \equiv \mathcal{U}_t$ does not depend on $x_t$, by the chain rule of differentiation, (5.25) for the required subgradient is extended to

$$\sum_{j=1}^{N} p_{tj} \left[\Psi'(y_{tj}, \bar{\theta}_t)\left(\nabla c_{tj}(x_t, \bar{u}_t) + A_{tj}^\top \nabla \underline{V}_{t+1}(A_{tj}x_t + B_{tj}\bar{u}_t + b_{tj})\right)\right], \tag{5.38}$$

where $\bar{u}_t$ and $\bar{\theta}_t$ are minimizers in the right hand side of (5.37), with $V_{t+1}(\cdot)$ replaced by its current approximation $\underline{V}_{t+1}(\cdot)$, and $\Psi'(y_{tj}, \bar{\theta}_t)$ is a subgradient of $\Psi(\cdot, \bar{\theta}_t)$ at $y_{tj} := c_{tj}(x_t, \bar{u}_t) + \underline{V}_{t+1}(A_{tj}x_t + B_{tj}\bar{u}_t + b_{tj})$.

The subgradient $\Psi'(\cdot, \theta)$ can be easily computed in many interesting examples (e.g., [68, Section 6.3.2]). For example, consider the convex combination of expectation and Average Value-at-Risk measure, defined in (4.6). The corresponding function $\Psi(z, \theta)$ of the form (5.36), is piecewise linear in $z$. Then $\Psi'(z, \theta) = 1 - \lambda_t$ if $z < \theta$, $\Psi'(z, \theta) = 1 - \lambda_t + \lambda_t(1 - \alpha_t)^{-1}$ if $z > \theta$; and if $z = \theta$, then $\Psi'(z, \theta)$ can be any point in the interval $[1 - \lambda_t, 1 - \lambda_t + \lambda_t(1 - \alpha_t)^{-1}]$.

The computed approximations $\underline{V}_{t+1}(\cdot)$ of the value functions define a feasible policy in the same way as in the risk neutral case. Also by construction we have that $V_t(\cdot) \geq \underline{V}_t(\cdot)$, $t = 1, ..., T$. Therefore the algorithm produces a (deterministic) lower bound for the optimal value of the problem.

Formula (5.37) shows that the controls and value of the parameter $\theta$ can be computed simultaneously. This leads to the following statistical upper bound for the (risk averse) value of the policy defined by the computed approximation. Let $\hat{\xi}_1, ..., \hat{\xi}_T$ be a sample path of the data process. Suppose for the moment that $\mathcal{R}_t = \mathbb{E}$, i.e., consider the risk neutral case. Let $\pi$ be a policy determined by controls $\bar{u}_t$ and states $\bar{x}_t$, and $\hat{c}_t := c_t(\bar{x}_t, \bar{u}_t, \hat{\xi}_t)$, $t = 1, ..., T$, and $\hat{c}_{T+1} := c_{T+1}(\bar{x}_{T+1})$, be cost values associated with the sample path. Then the corresponding point estimate of the expectation (3.1) of the total cost for the considered policy $\pi$ can be computed iteratively going backward in time with $\mathfrak{v}_{T+1} := \hat{c}_{T+1}$ and

$$\mathfrak{v}_t := \hat{c}_t + \mathfrak{v}_{t+1}, \quad t = T, ..., 1. \tag{5.39}$$

Value $\mathfrak{v}_1$ gives an unbiased estimate of the corresponding expected total cost.

This process can be adapted for general risk averse case as follows. Let $\hat{A}_t = A_t(\hat{\xi}_t)$, $\hat{B}_t = B_t(\hat{\xi}_t)$ and $\hat{b}_t = b_t(\hat{\xi}_t)$, $t = 1, ..., T$, be realizations of the parameters corresponding to the generated sample path. For the approximations $\underline{V}_t(\cdot)$ consider the corresponding policy with controls $\bar{u}_t$ and parameters $\bar{\theta}_t$ computed going forward, starting with initial value $\bar{x}_1$ and using (compare with (5.27)-(5.28))

$$(\bar{u}_t, \bar{\theta}_t) \in \operatorname*{arg\,min}_{u_t \in \mathcal{U}_t(\bar{x}_t), \theta \in \Theta} \sum_{j=1}^{N} p_{tj} \Psi\big(c_{tj}(\bar{x}_t, u_t)$$
$$+\underline{V}_{t+1}(A_{tj}\bar{x}_t + B_{tj}u_t + b_{tj}), \theta\big), \tag{5.40}$$
$$\bar{x}_{t+1} = \hat{A}_t \bar{x}_t + \hat{B}_t \bar{u}_t + \hat{b}_t. \tag{5.41}$$

Let $\hat{c}_t := c_t(\bar{x}_t, \bar{u}_t, \hat{\xi}_t)$, $t = 1, ..., T$, and $\hat{c}_{T+1} := c_{T+1}(\bar{x}_{T+1})$, be cost values associated with the generated sample path. Consider the following values defined iteratively going backward in time: $\mathfrak{v}_{T+1} := \hat{c}_{T+1}$ and

$$\mathfrak{v}_t := \Psi(\hat{c}_t + \mathfrak{v}_{t+1}, \bar{\theta}_t), \quad t = T, ..., 1. \tag{5.42}$$

It is possible to show that $\mathbb{E}[\mathfrak{v}_1]$ is greater than or equal to value of the policy defined by the considered approximate value functions, and hence is an upper bound on the optimal value of the risk averse problem (cf., [22]). Expectation $\mathbb{E}[\mathfrak{v}_1]$ can be estimated by randomly generating sample paths (scenarios) and averaging the computed realizations of random variable $\mathfrak{v}_1$. This is similar to the construction of a statistical upper bound in the risk neutral setting. The inequality: "$\mathbb{E}[\mathfrak{v}_1] \geq$ value of the constructed policy", is based on Jensen's inequality applied to convex function $\Psi(\cdot, \theta)$ and, unlike the risk neutral case, can be strict. Therefore the above procedure introduces an additional error in the estimated optimality gap. This additional error becomes larger when function $\Psi(\cdot, \theta)$ is "more nonlinear". Numerical experiments indicate that this procedure gives a reasonable upper bound, for instance for risk measures of the form (4.6) (cf., [22]).

**Q-factor approach.** Consider the dynamic programming equations (5.37) and define (see Remark 5.6)

$$Q_t(x_t, u_t, \theta_t) := \mathbb{E}_{\mathbb{P}_t}\left[\Psi(c_t(x_t, u_t, \xi_t) + V_{t+1}(A_t x_t + B_t u_t + b_t), \theta_t)\right].$$

$$(5.43)$$

We have that

$$V_t(x_t) = \inf_{u_t \in \mathcal{U}_t, \theta_t \in \Theta} Q_t(x_t, u_t, \theta_t),$$

and hence dynamic programming equations (5.37) can be written in terms of $Q_t(x_t, u_t, \theta_t)$ as

$$Q_t(x_t, u_t, \theta_t) = \mathbb{E}_{\mathbb{P}_t}[\Psi(c_t(x_t, u_t, \xi_t)$$
$$+ \inf_{u_{t+1} \in \mathcal{U}_{t+1}, \theta_{t+1} \in \Theta} Q_{t+1}(A_t x_t + B_t u_t + b_t, u_{t+1}, \theta_{t+1}), \theta_t)].$$

The cutting plane algorithm can be applied directly to functions $Q_t(x_t, u_t, \theta_t)$ rather than to the value functions $V_t(x_t)$. In the backward step of the algorithm, subgradients with respect to $x_t, u_t$ and $\theta_t$, of the current approximations of $Q_t(x_t, u_t, \theta_t)$ should be computed at the respective trial points, and the obtained cutting plane $\ell_t(x_t, u_t, \theta_t)$ is added to the current family of cutting planes of $Q_t(x_t, u_t, \theta_t)$. An advantage of that approach could be that the calculation of the respective subgradients does not require solving *nonlinear optimization* programs even if the function $\Psi$ is not polyhedral.

### 5.3.2 Infinite horizon setting

An infinite horizon (stationary) counterpart of the SOC problems (3.1)-(3.3) is

$$\min_{\pi \in \Pi} \mathbb{E}^{\pi} \left[ \sum_{t=1}^{\infty} \gamma^{t-1} c(x_t, u_t, \xi_t) \right], \quad (5.44)$$

where $\gamma \in (0,1)$ is the so-called discount factor. The optimization (minimization) in (5.44) is over the set of feasible policies $\Pi$ satisfying[8] (w.p.1):

$$u_t \in \mathcal{U}, \ x_{t+1} = F(x_t, u_t, \xi_t), \ t \geq 1. \quad (5.45)$$

It is assumed that $\xi_t \sim P$, $t = 1, ...,$ is an i.i.d. (independent identically distributed) sequence of random vectors with common distribution $P$. The cost $c(x, u, \xi)$ and the mapping $F(x, u, \xi)$ are assumed to be the same at every stage, and the (nonempty) feasibility set $\mathcal{U}$ of controls does not depend on either the state or stage.

The Bellman equation for the value function, associated with (5.44), can be written as

$$V(x) = \inf_{u \in \mathcal{U}} \mathbb{E}_P[c(x, u, \xi) + \gamma V(F(x, u, \xi))]. \quad (5.46)$$

Assuming that $c(x, u, \xi)$ is bounded, (5.46) has a unique solution $\bar{V}(\cdot)$ as the optimal value function. The optimal policy for (5.45) is given by $u_t = \pi(x_t)$, $t = 1, ...,$ with

$$\pi(\cdot) \in \arg\min_{u \in \mathcal{U}} \mathbb{E}_P[c(\cdot, u, \xi) + \gamma \bar{V}(F(\cdot, u, \xi))], \quad (5.47)$$

starting with initial value $x_1$ of the state vector.

Similar to Assumption 3.1, let us make following assumption ensuring convexity of the solution $\bar{V}(\cdot)$ of (5.46).

**Assumption 5.2.** The function $c(x, u, \xi)$ is convex in $(x, u)$, the mapping $F(x, u, \xi) = A(\xi)x + B(\xi)u + b(\xi)$ is affine, and the set $\mathcal{U}$ is convex closed.

---

[8]When the number of scenarios is finite, the feasibility constraints should be satisfied for all possible realizations of the random process $\xi_t$.

In order to numerically solve (5.46) we need to discretize the possibly continuous distribution $P$. Suppose that the SAA method is applied, i.e., a sample $\xi^1, ..., \xi^N$ of size $N$ from the distribution $P$ is generated (say by Monte Carlo sampling techniques), and the distribution $P$ in the Bellman equation (5.46) is replaced by its empirical estimate[9] $P_N = N^{-1} \sum_{j=1}^{N} \delta_{\xi^j}$, assigning probability $1/N$ to each generated point. This raises the question of the involved sample complexity, i.e., how large should be the sample size $N$ in order for the SAA problem to give an accurate approximation of the original problem. In some applications the discount factor is very close to one. It is well known that as the discount factor $\gamma$ approaches one, it becomes more difficult to solve the problem. Note that the optimal value of (5.44) increases at the rate of $O((1-\gamma)^{-1})$ as $\gamma$ approaches one. It is shown in [66] that the required sample size $N$ is of order $O((1-\gamma)^{-1}\varepsilon^{-2})$ as a function of the discount factor $\gamma$ and the error level $\varepsilon > 0$. That is, the error grows more or less proportionally to the optimal value as $\gamma$ approaches one. This indicates that the sample size required to achieve a specified *relative* error is not sensitive to the discount factor being close to one.

Suppose now that random vector $\xi$ has a finite number $\xi^1, ..., \xi^N$ of possible values with respective probabilities $p_1, ..., p_N$. For example, in the SAA approach this is a random sample generated by Monte Carlo sampling techniques, in which case $p_i = 1/N$, $i = 1, ..., N$. In that setting, the Bellman equation (5.46) can be written as

$$V(x) = \inf_{u \in \mathcal{U}} \sum_{j=1}^{N} p_j [c_j(x, u) + \gamma V(A_j x + B_j u + b_j)], \qquad (5.48)$$

where $c_j(x, u) := c(x, u, \xi^j)$, $A_j := A(\xi^j)$, $B_j := B(\xi^j)$ and $b_j := b(\xi^j)$, $j = 1, ..., N$.

A cutting plane algorithm, such as SDDP, can be applied to (5.48). Given a current approximation $\underline{V}(\cdot)$ of the value function by cutting planes and a trial point $\tilde{x}$, a new cutting plane $\ell(x) = \tilde{v} + \mathbf{g}^\top(x - \tilde{x})$ is added to the set of cutting planes of the value function, where

---

[9]By $\delta_\xi$ we denote measure assigning mass one at point $\xi$, the so-called Dirac measure.

$$\tilde{u} \in \arg\min_{u \in \mathcal{U}} \sum_{j=1}^{N} p_j [c_j(\tilde{x}, u) + \gamma \underline{V}(A_j \tilde{x} + B_j u + b_j)], \quad (5.49)$$

$$\tilde{v} = \sum_{j=1}^{N} p_j [c_j(\tilde{x}, \tilde{u}) + \gamma \underline{V}(A_j \tilde{x} + B_j \tilde{u} + b_j)], \quad (5.50)$$

$$g = \sum_{j=1}^{N} p_j [\nabla c_j(\tilde{x}, \tilde{u}) + \gamma A_j^\top \nabla \underline{V}(A_j \tilde{x} + B_j \tilde{u} + b_j)]. \quad (5.51)$$

If $c(x, u, \xi)$ is linear (polyhedral) in $u$ and the set $\mathcal{U}$ is polyhedral given by linear inequalities, the minimization problem in the right hand side of (5.49) can be formulated as a linear programming problem. The subgradient of $\underline{V}(\cdot)$ at a point $x$ can be computed in the same way as in (5.25) by using the subgradient of its supporting plane at $x$.

Similar to the case of finite horizon, by construction we have that $\bar{V}(\cdot) \geq \underline{V}(\cdot)$, and hence $\underline{V}(x_1)$ gives a (deterministic) lower bound for the optimal value of (5.44) with initial value $x_1$ of the state vector. Important questions, related to the above algorithm, is how to generate trial points and how to construct an upper bound for the optimal value of (5.44). Current implementation, which was tested in numerical experiments, is to apply the forward step of the cutting plane algorithm to a truncated version of (5.44). Note that

$$\left| \sum_{t=T+1}^{\infty} \gamma^{t-1} c(x_t, u_t, \xi_t) \right| \leq \sum_{t=T+1}^{\infty} \gamma^{t-1} |c(x_t, u_t, \xi_t)| \leq \frac{\kappa \gamma^T}{1 - \gamma}, \quad (5.52)$$

where $\kappa$ is an upper bound for $|c(x, u, \xi)|$. Therefore the suggestion is to choose the horizon $T$ such that $\kappa \gamma^T (1 - \gamma)^{-1} \leq \varepsilon$, where $\varepsilon > 0$ is a prescribed accuracy. Then the forward steps are applied to the finite horizon version of the problem using the current approximation $\underline{V}(\cdot)$ of the value function. One run of the forward step, applied to the truncated version, will generate a point estimate (up to accuracy $\varepsilon$) of value of the constructed policy, and $T - 1$ trial points. It is possible to choose, say at random, one of these trial points for construction of the cutting plane for the value function.

Such choice of the trial points may not be optimal. The subject of how to best construct trial points in the above algorithm has not yet been carefully investigated in the literature. Another problem with this procedure is that when the discount factor is close to one, the required horizon $T$ could be too large for a numerical implementation; recall that in order to construct the statistical upper bound the forward step should be run a reasonable number of times. In such cases, if the problem can be formulated as a linear program, deterministic upper bounds of the dual problem can be more efficient (cf., [67]).

**Risk averse case.** The infinite horizon risk neutral formulation (5.44) can be extended to the risk averse setting by replacing the expectation operator in the Bellman equation (5.46) with a coherent risk measure $\mathcal{R}$. As such, the risk averse counterpart of the Bellman equation (5.46) is given by

$$V(x) = \inf_{u \in \mathcal{U}} \mathcal{R}[c(x, u, \xi) + \gamma V(F(x, u, \xi)]. \tag{5.53}$$

This equation is the nested counterpart of the risk neutral problem (5.44),

$$\min_{\pi \in \Pi} \lim_{T \to \infty} \mathfrak{R}_T \left( \sum_{t=1}^{T} \gamma^{t-1} c(x_t, u_t, \xi_t) \right), \tag{5.54}$$

where $\mathfrak{R}_T$ is the corresponding $T$-stage nested risk measure (cf., [62]). As discussed in Section 3, for a considered policy, state $x_t$ and control $u_t$ at stage $t$ are functions of $\xi_{[t-1]}$. Consider the corresponding cost $Z_t := \gamma^{t-1} c(x_t, u_t, \xi_t)$ which is a function of $\xi_{[t]}$. Then

$$\mathfrak{R}_T \left( \sum_{t=1}^{T} Z_t \right) = \mathcal{R} \left( Z_1 + \mathcal{R}_{|\xi_{[1]}} (Z_2 + \dots + \mathcal{R}_{|\xi_{T-1}} (Z_T)) \right), \tag{5.55}$$

where $\mathcal{R}_{|\xi_{[t]}}$ is the conditional counterpart of $\mathcal{R}$.

The backward step of the cutting plane algorithm in the risk averse setting is basically the same as the one discussed above for the risk neutral case. In order to construct a statistical upper bound we can proceed similar to the construction of Section 5.3.1. That is, suppose that risk measure $\mathcal{R}$ is of the form (5.35) with function $\Psi(z, \theta)$ satisfying conditions (i) - (iii) specified in Section 5.3.1. Suppose also that $\xi$ has a finite number of possible values $\xi^1, \dots, \xi^N$ with reference probabilities $p_1, \dots, p_N$. Then the Bellman equation (5.53) can be written as

$$V(x) = \inf_{u \in \mathcal{U}, \theta \in \Theta} \sum_{j=1}^{N} p_j \Psi(c_j(x, u) + \gamma V(A_j x + B_j u + b_j), \theta). \tag{5.56}$$

A statistical upper bound can be constructed in a way similar to Section 5.3.1 using a truncated version of (5.44).

**Periodical setting** In various applications the data process $\xi_t$ has a periodical behavior. One such example is hydropower generation planning discussed in [71]. The uncertain process in that example is water inflows recorded on monthly basis. There is available historical data of 79 years of records of the natural monthly energy inflow. This allows one to model the uncertain process as a periodical random process with period $\mathsf{p} = 12$. The planning in that example was for 5 years with another 5 years to mitigate the end of horizon effect. This resulted in a $T = 120$ stage stochastic optimization problem. The periodical approach, which we discuss below, yields a drastic reduction in the number of stages while giving almost the same policy value for the first stage.

In Shapiro *et al.* [69], such periodical approach was introduced from the stochastic programming point of view. We discuss this in the framework of SOC. Consider the following infinite horizon SOC problem with discount factor $\gamma \in (0, 1)$:

$$\min_{u_t \in \mathcal{U}_t} \quad \mathbb{E}^\pi \left[ \sum_{t=1}^{\infty} \gamma^{t-1} c_t(x_t, u_t, \xi_t) \right] \tag{5.57}$$

$$\text{s.t.} \quad x_{t+1} = F_t(x_t, u_t, \xi_t), \ t \geq 1. \tag{5.58}$$

We assume the following periodic structure, with period $\mathsf{p} \geq 1$: the random data process $\xi_t$ is stagewise independent with $\xi_t$ and $\xi_{t+\mathsf{p}}$ having the same probability distribution and with $c_t(\cdot, \cdot, \cdot) = c_{t+\mathsf{p}}(\cdot, \cdot, \cdot)$, $F_t(\cdot, \cdot, \cdot) = F_{t+\mathsf{p}}(\cdot, \cdot, \cdot)$ and $\mathcal{U}_t = \mathcal{U}_{t+\mathsf{p}}$, for $t \geq 1$.

The Bellman equations for this problem take the form (compare with (3.11))

$$V_t(x_t) = \inf_{u_t \in \mathcal{U}_t} \mathbb{E} \left[ c_t(x_t, u_t, \xi_t) + \gamma V_{t+1}\big(F_t(x_t, u_t, \xi_t)\big) \right], \ t = 1, ..., \mathsf{p} - 1, \tag{5.59}$$

$$V_\mathsf{p}(x_\mathsf{p}) = \inf_{u_\mathsf{p} \in \mathcal{U}_\mathsf{p}} \mathbb{E} \left[ c_\mathsf{p}(x_\mathsf{p}, u_\mathsf{p}, \xi_\mathsf{p}) + \gamma V_1\big(F_\mathsf{p}(x_\mathsf{p}, u_\mathsf{p}, \xi_\mathsf{p})\big) \right], \ t = \mathsf{p}. \tag{5.60}$$

For period $\mathsf{p} = 1$, periodical problem (5.57)-(5.58) coincides with the stationary problem (5.44) of Section 5.3.2, and (5.60) becomes the Bellman equation (5.46). The optimal policy for (5.57)-(5.58) is given by $u_t = \pi_t(x_t)$ with

$$\pi_t(x_t) \in \inf_{u_t \in \mathcal{U}_t} \mathbb{E} \left[ c_t(x_t, u_t, \xi_t) + \gamma \bar{V}_{t+1}\big(F_t(x_t, u_t, \xi_t)\big) \right], \ t = 1, ..., \mathsf{p} - 1, \tag{5.61}$$

$$\pi_\mathsf{p}(x_\mathsf{p}) \in \inf_{u_\mathsf{p} \in \mathcal{U}_\mathsf{p}} \mathbb{E} \left[ c_\mathsf{p}(x_\mathsf{p}, u_\mathsf{p}, \xi_\mathsf{p}) + \gamma \bar{V}_1\big(F_\mathsf{p}(x_\mathsf{p}, u_\mathsf{p}, \xi_\mathsf{p})\big) \right], \ t = \mathsf{p}, \tag{5.62}$$

and $\pi_{t+p}(\cdot) = \pi_t(\cdot)$ for $t \geq 1$, where $(\bar{V}_1, ..., \bar{V}_p)$ is the solution of (5.59)-(5.60).

Such periodical formulation can be extended to the (nested) risk averse setting, and cutting plane type algorithms can be applied in a way similar to what was discussed above (cf., [69]).

# 6

# Computational Complexity of Cutting Plane Methods

As mentioned earlier, SDDP reduces to Kelly's cutting plane method for two stage problems and it can behave poorly as the number of decision variables increases. On the other hand, numerical experiments indicate that this type of algorithm can scale well with respect to the number of stages $T$. In this section we summarize some recent theoretical studies in [25], [31] on the rate of convergence of SDDP. We first establish the number of iterations required by the SDDP method and then present a variant of SDDP that can improve this complexity bound in terms of the dependence on the number of stages. Related developments, with a focus more on mixed-integer nonlinear optimization, can also be found in [74].

## 6.1 Computational Complexity of SDDP for SP

Consider the multistage stochastic programming problem (2.1)-(2.2). Similar to the previous section, we assume that $\xi_t$'s are stagewise independent (i.e., Assumption 2.2) and finitely supported (i.e., Assumption 5.1). For simplicity, we further assume that the probabilities for $\xi_{ti}$, $i = 1, \ldots, N$, are equal, i.e., $p_{ti} = 1/N$, $t = 2, ..., T$, $i = 1, ..., N$. For convenience, we denote the data used in the first stage by $\xi_{11} = \xi_1$. In that case, the dynamic programming equations in (2.8)-(2.9) reduce to

$$Q_t(x_{t-1}, \xi_{ti}) = \inf_{x_t \in \mathcal{X}_t} \{F_{ti}(x_t) := f_t(x_t, \xi_{ti}) + \mathcal{Q}_{t+1}(x_t) : B_{ti}x_{t-1} + A_{ti}x_t = b_{ti}\},$$

(6.1)

where

$$\mathcal{Q}_{t+1}(x_t) = \tfrac{1}{N}\textstyle\sum_{i=1}^{N} Q_{t+1}(x_t, \xi_{t+1,i}).$$

(6.2)

Our goal is to provide guarantees for the optimality gap of the solution produced by the SDDP method applied to (6.1). Observe that if the random variables $\xi_t$ in (2.1)-(2.2) are continuous, one needs to account for the approximation error caused by using Monte Carlo sampling techniques (see Remark 5.2), in addition to the suboptimality considered in this section for the SDDP method applied to the SAA problem.

The basic scheme of the SDDP method has been discussed in Section 5.1. Here we add a little more details to facilitate its complexity analysis. Each iteration of the SDDP method to solve (2.1)-(2.2) contains a forward step and a backward step. The forward step is the construction of a trajectory where the decisions are chosen according to the current approximation of value functions, while the backward step refers to the entire backward recursion to update the approximation of value functions. More specifically, in the forward step of the SDDP method, we randomly pick up an index $i_t$ out of $\{1, \ldots, N\}$ and solve problem

$$x_t^k \in \text{arg min} \left\{ \underline{F}_{ti_t}^{k-1}(x_t) := f_t(x_t, \xi_{ti_t}) + \underline{\mathcal{Q}}_{t+1}^{k-1}(x_t) \right\}$$
$$\text{s.t.} \quad B_{ti_t}x_{t-1}^k + A_{ti_t}x_t = b_{ti_t}, \ x_t \in \mathcal{X}_t,$$

to update $x_t^k$, which will be used as a trial point for updating $\underline{\mathcal{Q}}_{t+1}^{k-1}(\cdot)$ in the backward step. Here, we use a unified notation $x_t^k$ to denote $\bar{x}_t$ and $\tilde{x}_t$, which were introduced in the generic description of the SDDP method (see Remark 5.4). Equivalently, one can view $x_t^k$ as being chosen uniformly at random from $\tilde{x}_{ti}^k$, $i = 1, \ldots, N$, defined as

$$\tilde{x}_{ti}^k \in \text{arg min} \left\{ \underline{F}_{ti}^{k-1}(x_t) := f_t(x_t, \xi_{ti}) + \underline{\mathcal{Q}}_{t+1}^{k-1}(x_t) \right\}$$
$$\text{s.t.} \quad B_{ti}x_{t-1}^k + A_{ti}x_t = b_{ti}, \ x_t \in \mathcal{X}_t.$$

(6.3)

Note that we do not need to compute $\tilde{x}_{ti}^k$ for $i \neq i_t$, even though they will be used in the analysis of the SDDP method. It turns out that the way to select trial points directly impacts the rate of convergence of dynamic cutting plane methods. In the next subsection, we will present a variant of SDDP which chooses the trial points $x_t^k$ in a more aggressive manner to improve the rate of convergence.

In the backward step of SDDP, starting from the last stage, we update the cutting plane models $\underline{\mathcal{Q}}_t^k$ according to

$$\underline{\mathcal{Q}}_t^k(x_{t-1}) = \max\left\{\underline{\mathcal{Q}}_t^{k-1}(x_{t-1}), \tfrac{1}{N}\textstyle\sum_{i=1}^{N}\left[\mathfrak{Q}_{ti}^k(x_{t-1}^k) + \langle\nabla\mathfrak{Q}_{ti}^k(x_{t-1}^k), x_{t-1} - x_{t-1}^k\rangle\right]\right\},$$

for $t = T, T-1, \ldots, 2$, with $\underline{\mathcal{Q}}_t^0(x) = -\infty$. Here $\mathfrak{Q}_{ti}^k(x_{t-1}^k)$ and $\nabla\mathfrak{Q}_{ti}^k$ $(x_{t-1}^k)$, respectively, denote the optimal value and a subgradient at the point $x_{t-1}^k$ for the following approximate value function

$$\mathfrak{Q}_{ti}^k(x_{t-1}^k) = \min\left\{\underline{F}_{ti}^k(x_t) := f_t(x_t, \xi_{ti}) + \underline{\mathcal{Q}}_{t+1}^k(x_t)\right\}$$
$$\text{s.t. } B_{ti}x_{t-1}^k + A_{ti}x_t = b_{ti}, x_t \in \mathcal{X}_t.$$

In order to assess the progress made by each SDDP iteration, we need to introduce a few notions. We say that a search point $x_t^k$ is $\epsilon_t$-*saturated* at iteration $k$ if

$$\mathcal{Q}_{t+1}(x_t^k) - \underline{\mathcal{Q}}_{t+1}^k(x_t^k) \le \epsilon_t.$$

Intuitively, a point $x_t^k$ is saturated if we have a good approximation for the value function $\mathfrak{Q}_{t+1}$ at $x_t^k$. Moreover, we say an $\epsilon_t$-saturated search point $x_t^k$ at stage $t$ is $\delta_t$-*distinguishable* if $\|x_t^k - x_t^j\| > \delta_t$ for all other $\epsilon_t$-saturated search points $x_t^j$ for some $j \le k-1$, that have been generated for stage $t$ so far by the algorithm. Equivalently, an $\epsilon_t$-saturated search point $x_t^k$ is $\delta_t$-distinguishable if $\text{dist}(x_t^k, S_t^{k-1}) > \delta_t$, where $S_t^{k-1}$ the set of saturated points at stage $t$. Recall that $\text{dist}(x, S)$ denotes the distance from point $x$ to the set $S$. Hence, in this case, this search point $x_t^k$ is far away enough from other saturated points. The rate of convergence of SDDP depends on how fast it generates $\epsilon_t$-saturated and $\delta_t$-distinguishable search points. For simplicity, we assume in this work that $\epsilon_t = \delta_t = \epsilon$, $t = 1, \ldots, T$, for a given tolerance $\epsilon > 0$.

Under certain regularity assumptions, it can be shown that the function $F_{ti}$ and lower approximation functions $\underline{F}_{ti}^k$ are Lipschitz continuous modulus $M$. Note that $M$ depends on problem data but not on the particular algorithm. Utilizing these continuity assumptions, it can be shown that the probability for SDDP to find a new $\epsilon$-distinguishable and $\epsilon$-saturated search point at the $k$-th iteration can be bounded from below by

$$\tfrac{1}{\bar{N}}(1 - \Pr\{\tilde{g}_t^k \le \epsilon, t = 1, \ldots, T-1\}), \tag{6.4}$$

where

$$\bar{N} := N^{T-2} \quad \text{and} \quad \tilde{g}_t^k := \tfrac{1}{N}\sum_{i=1}^{N} \mathrm{dist}(\tilde{x}_{ti}^k, S_t^{k-1}). \tag{6.5}$$

On the other hand, if for some iteration $k$, we have $\tilde{g}_t^k \le \epsilon$ for all $t = 1, \ldots, T-1$, then

$$F_{11}(x_1^k) - F^* \le F_{11}(x_1^k) - \underline{F}_{11}^{k-1}(x_1^k) \le 2M(T-1)\epsilon. \tag{6.6}$$

Note that the upper bound $2M(T-1)$ can be further reduced so that it does not depend on $T$ if a discounting factor is incorporated in each stage of the problem.

Now let $K$ denote the number of iterations performed by the SDDP method before it finds a forward path $(x_1^k, \ldots, x_T^k)$ such that (6.6) holds. Using the previous observation in (6.4), it can be shown (see [31]) that $\mathbb{E}[K] \le \bar{K}_\epsilon \bar{N} + 2$, where $\bar{N}$ is defined in (6.5) and

$$\bar{K}_\epsilon := (T-1)\left(\tfrac{D}{\epsilon} + 1\right)^n. \tag{6.7}$$

Here $n_t \le n$ and $D_t \le D$ for $t = 1, \ldots, T$ denote the dimension and diameter of the feasible sets $\mathcal{X}_t$, respectively. In addition, for any $\alpha \ge 1$, we have

$$\Pr\{K \ge \alpha \bar{K}_\epsilon \bar{N} + 1\} \le \exp\left(-\tfrac{(\alpha-1)^2 \bar{K}_\epsilon^2}{2\alpha N}\right).$$

We now add a few remarks about the above complexity results for SDDP. Firstly, since SDDP is a randomized algorithm, we provide bounds on the expected number of iterations required to find an approximate solution of (2.1)-(2.2). We also show that the probability of having large deviations from these expected bounds for SDDP decays exponentially fast. Secondly, the complexity bound for the SDDP method depends on $\bar{N}$, which increases exponentially with respect to $T$. We will show in the next section how to reduce the dependence on $T$ by presenting a variant of the SDDP method.

## 6.2 Explorative Dual Dynamic Programming for SP

The SDDP method in the previous section chooses the feasible solution (and trial point) $x_t^k$ in a randomized manner. In this section, we will introduce a deterministic method, called Explorative Dual Dynamic Programming (EDDP) [31], which chooses the trial points in an aggressive manner to update the cutting plane models. As we will see, the latter approach will exhibit better iteration complexity while the former one is easier to implement.

The EDDP method consists of the forward step and backward step. In the forward step of EDDP, for each stage $t$, we solve $N$ subproblems defined in (6.3) to compute the search points $\tilde{x}_{ti}^k$, $i = 1, \ldots, N$. For each $\tilde{x}_{ti}^k$, we further compute the quantity $\mathrm{dist}(\tilde{x}_{ti}^k, S_t^{k-1})$, the distance between $\tilde{x}_{ti}^k$ and the set $S_t^{k-1}$ of currently saturated search points in stage $t$. Then we will choose from $\tilde{x}_{ti}^k$, $i = 1, \ldots, N$, the one with the largest value of $\mathrm{dist}(\tilde{x}_{ti}^k, S_t^{k-1})$ as $x_t^k$, i.e., $\mathrm{dist}(x_t^k, S_t^{k-1}) = \max_{i=1,\ldots,N} \mathrm{dist}(\tilde{x}_{ti}^k, S_t^{k-1})$. We can break ties arbitrarily. In view of the above discussion, the EDDP method always chooses the most "distinguishable" forward path to encourage exploration in an aggressive manner (see Figure 6.1). This also explains the origin of the name EDDP. The backward step of EDDP is similar to SDDP.

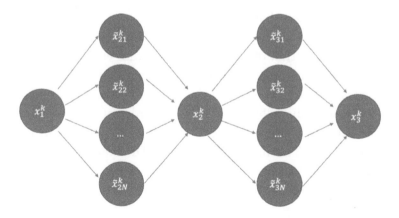

**Figure 6.1:** EDDP picks $x_2^k$, $x_3^k$, ..., in an aggressive manner to maximize $g_t^k(\tilde{x}_{ti}^k) := \min_{s \in S_t^{k-1}} \|s - \tilde{x}_{ti}^k\|$, where $S_t^{k-1}$ denotes the set of previously saturated points at stage $t$.

Under the same regularity assumptions as in SDDP, it can be shown that each EDDP iteration will either generate at least one $\epsilon$-

distinguishable and $\epsilon$-saturated search point at some stage $t = 1, \ldots, T$, or find a feasible solution $x_1^k$ of (2.1)-(2.2) such that

$$F_{11}(x_1^k) - F^* \leq 2M(T-1)\,\epsilon. \tag{6.8}$$

As a result, the total number of iterations performed by EDDP before finding a feasible policy of (2.1)-(2.2) satisfying (6.8) within at most $\bar{K}_\epsilon +$ 1 iterations where $\bar{K}_\epsilon$ is defined in (6.7) (see [31]). This complexity bound does not depend on the number of scenarios $\bar{N}$, and hence significantly outperforms the one for the original SDDP method. Moreover, we can show that the relation $\text{dist}(x_1^k, S_1^{k-1}) \leq \delta$ implies the relation in (6.8) for a properly chosen value of $\delta$, and hence it can be used as a termination criterion for the EDDP method. Note that it is possible to further reduce the dependence of the computational complexity of EDDP on $T$ so that the term $\bar{K}_\epsilon$ in (6.7) does not depend on $T$. However, this would require us to further modify EDDP by incorporating early termination of the forward step. More specifically, we need to terminate the forward step at some stage $\bar{t} \leq T$ and start the backward step from $\bar{t}$, whenever $\text{dist}(x_{\bar{t}}^k, S_{\bar{t}}^{k-1})$ falls within a certain threshold value (see [25], [74]).

It should be noted that although the complexity of SDDP is worse than that for EDDP, its performance in earlier steps of the algorithm should be similar to that of EDDP. Intuitively, for earlier iterations, the tolerance parameter $\delta_t$ used to define $\delta_t$-distinguishable points are large. As long as $\delta_t$ are large enough so that the solutions $\tilde{x}_{ti}^k$ are contained within a ball with diameter roughly in the order of $\delta_t$, one can choose any point randomly from $\tilde{x}_{ti}^k$ as $x_t^k$. In this case, SDDP will perform similarly to EDDP. This may explain why SDDP exhibits good practical performance for low accuracy region. For high accuracy region, the new EDDP algorithm seems to be a much better choice in terms of its theoretical complexity.

## 6.3   Complexity of SDDP and EDDP for SOC

The complexity analysis for SP introduced in the previous two subsections for SDDP and EDDP can be performed for SOC in a straightforward manner. In Section 5.2, two possible approaches have been discussed for SOC: one to approximate the functions $V_t(x_t)$, and the other to approximate the function $Q_t(x_t, u_t)$. The complexities of the SDDP and EDDP methods will be different for these two approaches. Specifically, the rate of convergence for the approach based on approximating $V_t(\cdot)$ does not depend on the dimensions of $u_t$'s, while the one based on approximating $Q_t(\cdot, \cdot)$ exponentially depends on the dimensions of $u_t$'s. Therefore, the former approach should be preferred to the latter one if the dimensions of control variables are large.

## 6.4 Complexity of SDDP and EDDP for Risk Averse Problems

We can extend the complexity analysis of EDDP to risk averse SP models as described in (5.33) in a straightforward way. The trial point $x_t^k$ used in EDDP always corresponds the most distinguishable one among all the scenarios $i = 1, \ldots, N$, and hence the supremum taken over $\zeta$ in (5.33) does not impact the saturation of search points. However, it is not clear yet how the complexity analysis of SDDP can be extended to this risk averse model. On the other hand, the complexity analysis of both SDDP and EDDP can be readily extended to the risk averse SOC models as described in (5.35).

## 6.5 Complexity of SDDP and EDDP over an Infinite Horizon

It is possible to generalize EDDP and SDDP to solve stochastic programs over an infinite horizon (see Section 5.3.2), when the number of stages $T = \infty$, the cost function $f_t = f$, the feasible set $X_t = X$ for all $t$, and the random variables $\xi_t$ are iid (independent identically distributed) and can be viewed as realizations of random vector $\xi$. A common line is to approximate the infinite horizon with a finite horizon. For example, using the relation (5.52) in the SOC problem, a horizon length of $T = O((\epsilon(1 - \gamma))^{-1})$ suffices for a target accuracy $\epsilon$. A direct extension of EDDP method show that the computational complexity depends only quadratically on the time horizon $T$. Exploiting the fact that the infinite-horizon problem is stationary, it has been shown that a simple modification that combines the forward and backward step can improve the dependence on $T$ from quadratic to linear. Alternative selection strategies, including the use of upper bounds on the cost-to-go function and random sampling as in SDDP can also be incorporated, see [25] for more detail as well as some numerical illustration for EDDP over an infinite horizon.

# 7

---

# Dynamic Stochastic Approximation Algorithms

---

## 7.1 Extension of Stochastic Approximation

Stochastic approximation (SA) has attracted much attention recently for solving static stochastic optimization problems given in the form of

$$\min_{x \in \mathcal{X}} \{f(x) := \mathbb{E}_\xi[F(x, \xi)]\}, \tag{7.1}$$

where $\mathcal{X}$ is a closed convex set, $\xi$ denotes the random vector and $F(\cdot, \xi)$ is a lower semicontinuous convex function. Observe that we can cast two-stage stochastic optimization in the form of (7.1) by viewing $F$ as the summation of the cost function of the first-stage problem and the value function of the second-stage problem (see [27], [33], [44]).

The basic SA algorithm, initially proposed by Robbins and Monro [57], mimics the simple projected gradient descent method by replacing exact gradient with its unbiased estimator. Important improvements for the SA methods have been made by Nemirovski and Yudin [45] and later by Polayk and Juditsky [54], [55]. During the past few years, significant progress has been made in SA methods including the incorporation of momentum and variance reduction, and generalization to nonconvex setting [29]. It has been shown in [33], [44] that SA type methods can significantly outperform the SAA approach for solving static (or two-stage) stochastic programming problems. However, it remains unclear whether these SA methods can be generalized for multistage stochastic optimization problems with $T \geq 3$.

In this section, we discuss a recently developed dynamic stochastic approximation (DSA) method for multistage stochastic optimization [34]. The basic idea of the DSA method is to apply an inexact primal-dual SA method for solving the $t$-th stage optimization problem to compute an approximate stochastic subgradient for its associated value functions. For simplicity, we focus on the basic scheme of the DSA algorithm and its main convergence properties for solving three-stage stochastic optimization problems. We also briefly discuss how to generalize it for solving more general form of multistage stochastic optimization with $T > 3$.

The main problem of interest in this section is the following three-stage SP problem:

$$
\begin{array}{lll}
\min f_1(x_1, \xi_1)+ & \mathbb{E}[\min\ f_2(x_2, \xi_2) & +\mathbb{E}_{|\xi_{[2]}}[\min\ f_3(x_3, \xi_3)]] \\
\text{s.t. } A_1 x_1 = b_1, & \text{s.t. } A_2 x_2 + B_2 x_1 = b_2, & \text{s.t. } A_3 x_3 + B_3 x_2 = b_3, \\
x_1 \in \mathcal{X}_1, & x_2 \in \mathcal{X}_2, & x_3 \in \mathcal{X}_3.
\end{array}
\tag{7.2}
$$

We can write (7.2) in a more compact form by using value functions as discussed in (2.3)-(2.5). More specifically, let $Q_3(x_2, \xi_{[3]})$ be the value function at the third stage and $\mathcal{Q}_3(x_2)$ be the corresponding expected value function conditionally on $\xi_{[2]}$:

$$
\begin{aligned}
Q_3(x_2, \xi_{[3]}) \quad &:= \quad \min\ f_3(x_3, \xi_3) \\
&\qquad \text{s.t. } A_3 x_3 + B_3 x_2 = b_3, \\
&\qquad\quad x_3 \in \mathcal{X}_3. \\
\mathcal{Q}_3(x_2, \xi_{[2]}) \quad &:= \quad \mathbb{E}_{|\xi_{[2]}}[Q_3(x_2, \xi_{[3]})].
\end{aligned}
\tag{7.3}
$$

We can then define the stochastic value function $Q_2(x_1, \xi_{[2]})$ and its corresponding (expected) value function as

$$
\begin{aligned}
Q_2(x_1, \xi_{[2]}) \quad &:= \quad \min\ f_2(x_2, \xi_2) + \mathcal{Q}_3(x_2, \xi_{[2]}) \\
&\qquad \text{s.t. } A_2 x_2 + B_2 x_1 = b_2, \\
&\qquad\quad x_2 \in \mathcal{X}_2. \\
\mathcal{Q}_2(x_1, \xi_1) \quad &:= \quad \mathbb{E}_{|\xi_1}[Q_2(x_1, \xi_{[2]})] = \mathbb{E}[Q_2(x_1, \xi_2)].
\end{aligned}
\tag{7.4}
$$

(7.2) can then be formulated equivalently as

$$\min \ f_1(x_1, \xi_1) + Q_2(x_1, \xi_1)$$
$$\text{s.t.} \ A_1 x_1 = b_1, \tag{7.5}$$
$$x_1 \in \mathcal{X}_1.$$

## 7.2 Approximate Stochastic Subgradients

In order to solve problem (7.5) by stochastic approximation type methods, we need to understand how to compute first-order information of the value functions $Q_2$ and $Q_3$. Recall that for a given closed convex set $\mathcal{X} \subseteq \mathbb{R}^n$ and a closed convex function $Q : \mathcal{X} \to \mathbb{R}$, $g(x)$ is called an $\epsilon$-subgradient of $Q$ at $x \in \mathcal{X}$ if

$$Q(y) \geq Q(x) + \langle g(x), y - x \rangle - \epsilon \quad \forall y \in \mathcal{X}. \tag{7.6}$$

The collection of all such $\epsilon$-subgradients of $Q$ at $x$ is called the $\epsilon$-subdifferential of $Q$ at $x$, denoted by $\partial_\epsilon Q(x)$. Since both $Q_2$ and $Q_3$ are given in the form of (conditional) expectation, their exact first-order information is hard to compute. We resort to the computation of a stochastic $\epsilon$-subgradient of these value functions defined as follows. Observe that we do not assume the stagewise independence throughout this section.

**Definition 7.1.** $G(x, \xi_{[t]})$ is called a stochastic $\epsilon$-subgradient of the value function $Q_t(x, \xi_{[t-1]}) = \mathbb{E}_{|\xi_{[t-1]}}[Q_t(x, \xi_{[t]})]$ if $G(x, \xi_{[t]})$ is an unbiased estimator of an $\epsilon$-subgradient of $Q_t(x, \xi_{[t-1]})$ with respect to $x$, i.e.,

$$\mathbb{E}_{|\xi_{[t-1]}}[G(x, \xi_{[t]})] = g(x, \xi_{[t-1]}) \quad \text{and} \quad g(x, \xi_{[t-1]}) \in \partial_\epsilon Q_t(x, \xi_{[t-1]}). \tag{7.7}$$

To compute a stochastic $\epsilon$-subgradient of $Q_2$ (resp., $Q_3$), we have to compute an approximate subgradient of the corresponding stochastic value function $Q_2(x_1, \xi_{[2]})$ (resp., $Q_3(x_2, \xi_{[3]})$). To this end, we further assume that strong Lagrange duality holds for the optimization problems defined in (7.4) (resp., (7.3)) almost surely. In other words, these problems can be formulated as saddle point problems:

$$Q_2(x_1, \xi_{[2]}) = \max_{y_2} \min_{x_2 \in \mathcal{X}_2} \langle b_2 - B_2 x_1 - A_2 x_2, y_2 \rangle + f_2(x_2, \xi_2) + Q_3(x_2, \xi_{[2]}),$$
$$\tag{7.8}$$

$$Q_3(x_2, \xi_{[3]}) = \max_{y_3} \min_{x_3 \in \mathcal{X}_3} \langle b_3 - B_3 x_2 - A_3 x_3, y_3 \rangle + f_3(x_3, \xi_3). \tag{7.9}$$

One sufficient condition to guarantee the equivalence between (7.4) (resp., (7.3)) and (7.8) (resp., (7.9)) is that (7.4) (resp., (7.3)) is solvable.

Observe that (7.8) and (7.9) are special cases of a more generic saddle point problem:

$$Q(u, \xi) := \max_{y} \min_{x \in \mathcal{X}} \langle b - Bu - Ax, y \rangle + f(x, \xi) + \mathcal{Q}(x), \qquad (7.10)$$

where $b$, $A$ and $B$ are linear mappings of the random variable $\xi$. For example, (7.9) is a special case of (7.10) with $u = x_2$, $y = y_3$, $\xi = \xi_3$, $f = f_3$ and $\mathcal{Q} = 0$. It is worth noting that the first stage problem can also be viewed as a special case of (7.10), since (7.5) is equivalent to

$$\max_{y_1} \min_{x_1 \in \mathcal{X}_1} \left\{ \langle b_1 - A_1 x_1, y_1 \rangle + f_1(x_1, \xi_1) + \mathcal{Q}_2(x_1, \xi_1) \right\}. \qquad (7.11)$$

We now discuss how to relate an approximate solution of the saddle point problem in (7.10) to an approximate subgradient of $Q$. Let $(x_*, y_*)$ be a pair of optimal solutions of the saddle point (7.10), i.e.,

$$Q(u, \xi) = \langle y_*, b - Bu - Ax_* \rangle + f(x_*, \xi) + \mathcal{Q}(x_*) = f(x_*, \xi) + \mathcal{Q}(x_*), \qquad (7.12)$$

where the second identity follows from the complementary slackness of Lagrange duality. It can be shown (see Lemma 1 of [34]) that if the gap function satisfies

$$\begin{aligned} \mathrm{gap}(\bar{z}; x, y_*) := {} & \langle y_*, b - Bu - A\bar{x} \rangle + f(\bar{x}, \xi) + \mathcal{Q}(\bar{x}) \\ & - \langle \bar{y}, b - Bu - Ax \rangle - f(x, \xi) - \mathcal{Q}(x) \leq \epsilon, \ \forall x \in \mathcal{X}, \end{aligned} \qquad (7.13)$$

for a given $\bar{z} := (\bar{x}, \bar{y})$ and $u$, then $-B^T \bar{y}$ is an $\epsilon$-subgradient of $Q(u, \xi)$ at $u$.

In view of this observation, in order to compute a stochastic subgradient of $\mathcal{Q}_t(u, \xi_{[t-1]}) = \mathbb{E}_{\xi_{[t-1]}}[Q_t(u, \xi_{[t]})]$ at a given point $u$, we can first generate a random realization $\xi_t$ conditionally on $\xi_{[t-1]}$ and then try to find a pair of solutions $(\bar{x}, \bar{y})$ satisfying

$$\begin{aligned} & \langle y_{t*}, b_t - B_t u - A_t \bar{x} \rangle + f(\bar{x}, \xi_t) + \mathcal{Q}_{t+1}(\bar{x}, \xi_{[t]}) \\ & - \langle \bar{y}, b_t - B_t u - A_t x \rangle - f(x, \xi_t) - \mathcal{Q}_{t+1}(x, \xi_{[t]}) \leq \epsilon, \ \forall x \in \mathcal{X}_t, \end{aligned}$$

where $y_{t*} \equiv y_{t*}(\xi_{[t]})$ denotes the optimal solution for the $t$-th stage problem associated with the random realization $\xi_{[t]}$. We will then use

$-B^T \bar{y}$ as a stochastic $\epsilon$-subgradient of $\mathcal{Q}_t(u, \xi_{[t-1]})$ at $u$. The difficulty exists in that the function $\mathcal{Q}_{t+1}(\bar{x}, \xi_{[t]})$ is also given in the form of expectation, and requires a numerical procedure to estimate its value. We will discuss this in more details in the next section.

## 7.3   The DSA Algorithm and its Convergence Properties

Our goal in this section is to present the basic scheme of the dynamic stochastic approximation algorithm applied to (7.5). This algorithm relies on the following three key primal-dual steps, referred to as stochastic primal-dual transformation (SPDT), applied to the generic saddle point problem in (7.10) at every stage.

$$(p_+, d_+, \tilde{d}) = \text{SPDT}(p, d, d\_, \mathcal{Q}', u, \xi, f, X, \theta, \tau, \eta):$$

$$\tilde{d} = \theta(d - d\_) + d. \tag{7.14}$$

$$p_+ = \arg\min_{x \in X} \langle b - Bu - Ax, \tilde{d} \rangle + f(x, \xi) + \langle \mathcal{Q}', x \rangle + \tfrac{\tau}{2}\|x - p\|^2. \tag{7.15}$$

$$d_+ = \arg\min_{y} \langle -b + Bu + Ap_+, y \rangle + \tfrac{\eta}{2}\|y - d\|^2. \tag{7.16}$$

In the above primal-dual transformation, the input $(p, d, d\_)$ denotes the primal solution in the current iteration, the dual solution in the current iteration, and the dual solution in the previous iteration, respectively. Moreover, the input $\mathcal{Q}'$ denotes a stochastic $\epsilon$-subgradient for $\mathcal{Q}$ at the current search point $p$. The parameters $(u, \xi, f, X)$ describe the problem in (7.10) and $(\theta, \tau, \eta)$ are certain algorithmic parameters to be specified. Given these input parameters, the relation in (7.14) defines a dual extrapolation (or prediction) step to estimate the dual variable $\tilde{d}$ for the next iterate. Based on this estimate, (7.15) performs a primal projection to compute $p_+$, and then (7.16) updates in the dual space to compute $d_+$ by using the updated $p_+$. Note that the Euclidean projection in (7.15) can be extended to the non-Euclidean setting by replacing the term $\|x - p\|^2/2$ with Bregman divergence. We assume that the above SPDT operator can be performed very fast or even has explicit expressions. The primal-dual transformation is closely related to the alternating direction method of multipliers and the primal-dual hybrid gradient method (see [34] for an account of history for this procedure).

In order to solve (7.5), we will combine the above primal-dual transformation applied to all the three stages, the scenario generation for the random variables $\xi_2$ and $\xi_3$ in the second and third stage, and certain averaging steps in both the primal and dual spaces to compute an approximate pair of primal and dual solution for the saddle point problem in the form of (7.10) at each stage (see Figure 7.1 for an illustration). More specifically, the DSA method consists of three loops. The innermost (third) loop runs $N_3$ steps of SPDT in order to compute an approximate stochastic subgradient of the value function $\mathcal{Q}_3$ of the third stage. The second loop consists of $N_2$ SPDTs applied to the saddle point formulation of the second-stage problem, which requires the output from the third loop. The outer loop applies $N_1$ SPDTs to the saddle point formulation of the first-stage optimization problem in (7.5), using the approximate stochastic subgradients for $\mathcal{Q}_2$ computed by the second loop. In this algorithm, we need to generate $N_1$ and $N_1 \times N_2$ realizations for the random vectors $\xi_2$ and $\xi_3$, respectively.

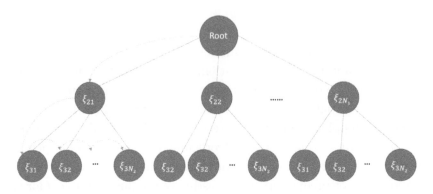

**Figure 7.1:** DSA performs a depth-first search through the scenario tree.

Observe that the DSA algorithm described above is conceptual only since we have not specified any algorithmic parameters $(\theta, \tau, \eta)$ yet. Two sets of parameters will be chosen depending on the stages where SPDTs are applied. One set of parameters that may lead to slightly slower rate of convergence but can guarantee the boundedness of the generated dual iterates will be used for the second and third stage, while another set of parameter setting is used in the first stage to achieve faster rate of convergence. Suppose that $f_t$, $t = 1, 2, 3$, are generally (not necessarily strongly) convex. By properly specifying the algorithmic parameters, it can be shown that by setting $N_3 = \mathcal{O}(1/\epsilon^2)$ and $N_2 = \mathcal{O}(1/\epsilon^2)$, we will find an approximate $\epsilon$-solution, i.e., a point $\bar{x}_1 \in \mathcal{X}_1$ such that

$$\mathbb{E}[f_1(\bar{x}_1, \xi_1) + \mathcal{Q}_2(\bar{x}_1, \xi_1) - (f_1(x_*, \xi_1) + \mathcal{Q}_2(x_*, \xi_1))] \leq \epsilon,$$
$$\mathbb{E}[\|A\bar{x}_1 - b\|] \leq \epsilon,$$

in at most $N_1 = \mathcal{O}(1/\epsilon^2)$ outer iterations. As a consequence, the number of random samples $\xi^2$ and $\xi^3$ used in the DSA method are bounded by

$$N_1 = \mathcal{O}(1/\epsilon^2) \quad \text{and} \quad N_1 \times N_2 = \mathcal{O}(1/\epsilon^4), \tag{7.17}$$

respectively. Now consider the case when the objective functions $f_t$, $t = 1, 2, 3$, are strongly convex. In that case the above complexity results can be significantly improved. In particular, by choosing $N_3 = \mathcal{O}(1/\sqrt{\epsilon})$ and $N_2 = \mathcal{O}(1/\epsilon)$, we will find an approximate $\epsilon$-solution in at most $N_1 = \mathcal{O}(1/\epsilon)$ outer iterations. As a result, it can be shown that the number of random samples of $\xi_2$ and $\xi_3$ will be bounded by $N_1$ and $N_1 \times N_2$, i.e., $\mathcal{O}(1/\epsilon)$ and $\mathcal{O}(1/\epsilon^2)$, respectively,

It should be noted that our analysis of DSA focuses on the optimality of the first-stage decisions, and the decisions we generated for the later stages are mainly used for computing the approximate stochastic subgradients for the value functions at each stage. Except for the first stage decision $\bar{x}_1$, the performance guarantees (e.g., feasibility and optimality) that we can provide for later stages are dependent on the sequences of random variables (or scenarios) we generated.

## 7.4 DSA for General Multistage Stochastic Optimization

In this section, we consider a multistage stochastic optimization problem given by

$$\begin{aligned} \min \quad & f_1(x_1, \xi_1) + \mathcal{Q}_2(x_1, \xi_1) \\ \text{s.t.} \quad & A_1 x_1 = b_1, \\ & x_1 \in \mathcal{X}_1, \end{aligned} \tag{7.18}$$

where the value factions $\mathcal{Q}_t$, $t = 2, \ldots, T$, are recursively defined by

$$
\begin{aligned}
\mathcal{Q}_t(x_{t-1}, \xi_{[t-1]}) \quad &:= F_{t-1}(x_{t-1}, \xi_{t-1}) + \mathbb{E}_{|\xi_{[t-1]}}[Q_t(x_{t-1}, \xi_{[t]})], \quad t = 2, \ldots, T-1, \\
Q_t(x_{t-1}, \xi_{[t]}) \quad &:= \min \ f_t(x_t, \xi_t) + \mathcal{Q}_{t+1}(x_t) \\
&\quad \text{s.t.} \ \ A_t x_t + B_t x_{t-1} = b_t, \\
&\quad\quad\quad\ x_t \in \mathcal{X}_t,
\end{aligned}
\tag{7.19}
$$

and

$$
\begin{aligned}
\mathcal{Q}_T(x_{T-1}, \xi_{[T-1]}) \quad &:= \mathbb{E}_{\xi_T | \xi_{[T-1]}}[Q_T(x_{T-1}, \xi_{[T]})], \\
Q_T(x_{T-1}, \xi_{[T]}) \quad &:= \min \ f_T(x_T, \xi_T) \\
&\quad \text{s.t.} \ \ A_T x_T + B_T x_{T-1} = b_T, \\
&\quad\quad\quad\ x_T \in \mathcal{X}_T.
\end{aligned}
\tag{7.20}
$$

Here $\xi_t$ are random variables, $f_t(\cdot, \xi_t)$ are relatively simple functions, and $F_t(\cdot, \xi_t)$ are general (not necessarily simple) Lipschitz continuous convex functions. We also assume that one can compute the subgradient $F'(x_t, \xi_t)$ of function $F_t(x_t, \xi_t)$ at any point $x_t \in \mathcal{X}_t$ for a given $\xi_t$.

(7.18) is more general than (7.2) (or equivalently, (7.5)) in the following sense. First, we are dealing with a more complicated multistage stochastic optimization problem where the number of stages $T$ in (7.18) can be greater than three. Second, the value function $\mathcal{Q}_t(x_{t-1}, \xi_{[t-1]})$ in (7.19) is defined as the summation of $F_{t-1}(x_{t-1}, \xi_{t-1})$ and $\mathbb{E}_{|\xi_{[t-1]}}[Q_t(x_{t-1}, \xi_{[t]})]$, where $F_{t-1}$ is not necessarily simple. The DSA algorithm can be generalized for solving (7.18). More specifically, we will call the SPDT operators to compute a stochastic $\epsilon$-subgradient of $\mathcal{Q}_{t+1}$ at $x_t$, $t = 1, \ldots, T-2$, in a recursive manner until we obtain the $\epsilon$-subgradient of $\mathcal{Q}_T$ at $x_{T-1}$.

Similar to the three-stage problem, let $N_t$ be the number of iterations for stage $t$ subproblem. For the last stage $T$, we will set $N_T = \mathcal{O}(1/\epsilon)$ and $N_T = \mathcal{O}(1/\sqrt{\epsilon})$ for the generally convex and strongly convex cases, respectively. For the middle stages $t = 2, \ldots, T-1$, we set $N_t = \mathcal{O}(1/\epsilon^2)$ and $N_t = \mathcal{O}(1/\epsilon)$. Different algorithmic settings will be used for either generally convex or strongly convex cases. Moreover, less aggressive stepsize selection will be used for the inner loops to guarantee the boundedness of the generated dual variables. Under these parameter selection, we can show that the number of outer loops performed by the DSA method to find an approximate $\epsilon$-solution can be bounded by $N_1 = \mathcal{O}(1/\epsilon^2)$ and $\mathcal{O}(1/\epsilon)$, respectively, for the generally convex and strongly convex problems.

In view of these results, the total number of scenarios required to find an $\epsilon$-solution of (7.18) is given by $N_2 \times N_3 \times \ldots N_T$, and hence will grow exponentially with respect to $T$, no matter whether the objective functions are strongly convex or not. These sampling complexity bounds match well with those lower bounds in [64], [70], implying that multistage stochastic optimization problems are essentially intractable for $T \geq 5$ and a moderate target accuracy. Hence, it is reasonable to use the DSA algorithm only for multistage stochastic optimization problems with $T = 3$ or 4 and $\epsilon$ relatively large. However, it is interesting to point out that the DSA algorithm only needs to go through the scenario tree once and hence its memory requirement increases only linearly with respect to $T$. Interested readers are also referred to [34] for some numerical illustration on the comparison between DSA and SDDP.

## 7.5 DSA for SOC Problems

Whenever SOC can be formulated as a convex multistage SP problem, it is possible to specialize or extend the DSA algorithm to solve it. In comparison with SDDP/EDDP methods, DSA can handle SOC problems with a large number of state and/or control variables. However, its complexity grows exponentially with the number of decision periods.

## 7.6 Combined EDDP and DSA for Hierarchical Problems

In this section, we discuss how EDDP/SDDP type algorithms can be applied together with DSA type algorithms for solving a class of hierarchical stationary stochastic programs (HSSPs). HSSPs can model problems with a hierarchy of decision-making, e.g., how managerial decisions influence day-to-day operations in a factory. More specifically, the upper level decisions are made by solving a stationary stochastic program over an infinite horizon (see Section 5.3.2), where the number of stages $T = \infty$, the cost function $f_t = f$, the feasible set $\mathcal{X}_t = \mathcal{X}$, and the random variables $\xi_t$ are iid having the same distribution as $\xi$. The lower level decisions are determined by a stochastic two-stage program,

$$f(x, \xi) := \min_{z_1 \in Z_1(x, \xi)} f_1(z_1, \xi) + \mathbb{E}_{\zeta}\Big[ \min_{z_2 \in Z_2(z_1, \zeta)} f_2(z_2, \zeta) \Big], \qquad (7.21)$$

where $\zeta$ denotes the random variable independent of $\xi$ involved in the second stage problem.

We extend the dual dynamic programming method to a so-called *hierarchical dual dynamic programming* by accounting for inexact solutions to the subproblems. To solve the lower-level stochastic multistage problem, we approximate it using SAA approach and solve the resulting problem using a primal-dual stochastic approximation method. This method can be generalized to a DSA-type method when the number of stages is three or more. Our results show that when solving an infinite-horizon hierarchical problem where the top-level decision variable is of modest size (i.e., $n = 4$) and the lower-level stochastic multistage program has a modest number of stages (2 or 3), then by integrating dual dynamic programming to handle the top-level decisions and stochastic approximation-type method for the lower-level decisions, we can get a polynomial computational complexity that is independent of the dimension of the lower-level decision variables (see [25] for detail).

# 8

## Conclusions

In this monograph, we review a few numerical methods, including cutting plane and SA approaches, for solving multistage stochastic optimization problems. By studying the performance guarantees for both types of methods, we conclude that cutting plane methods are more advantageous in dealing with problems with a large number of stages but a small number of state variables. On the other hand, SA approaches can handle more efficiently problems with a large number of state variables but a small number of stages. These two types of methods appear to be complementary to each other. Possible future research could be directed to how to combine the advantages of these methods, and some initial effort along this direction has been taken in [25].

Adding cutting planes increases computational complexity in the forward steps of the SDDP type algorithms. During the iteration process some of these cutting planes become redundant and could be removed. There is a trade off between computational time of identifying the redundant cutting planes and improving the computational complexity of the forward steps. Different strategies were suggested in the literature for computationally efficient procedures in that respect (e.g., [20], [39], [40]).

In this work, we discussed cutting plane methods applied to *convex* multistage stochastic programs. Many important applications involve integer (binary) decision variables. Such (even linear) mixed - integer problems in the multistage setting are notoriously difficult to solve and little progress has been reported so far. In that respect, we can mention recent papers [50], [74], [77]. In [77] the problem was approached by using Lagrangian cuts, derived from a Lagrangian relaxation, combined with a variant of the SDDP method. This so-called stochastic dual dynamic integer programming algorithm is included as an option in open-source libraries [11], [13].

For stochastic approximation type methods, we focused on risk neutral multistage stochastic programs. Recently there has been some progress in designing optimal SA methods for static convex risk averse problems [17], [73], [75]. It would be interesting if one can generalize these SA methods for convex risk averse multistage SP problems.

# Acknowledgements

The research of Guanghui Lan was partially supported by the NSF grant DMS-1953199 and the NSF AI Institute grant NSF-2112533.

The research of Alexander Shapiro was partially supported by Air Force Office of Scientific Research (AFOSR) under Grant FA9550-22-1-0244.

# References

[1] P. Artzner, F. Delbaen, J.-M. Eber, and D. Heath, "Coherent measures of risk," *Mathematical Finance*, vol. 9, 1999, pp. 203–228.

[2] E. M. L. Beale, "On minimizing a convex function subject to linear inequalities," *Journal of the Royal Statistical Society, Series B*, vol. 17, 1955, pp. 173–184.

[3] R. Bellman, *Dynamic Programming*. Princton, NJ: Princeton University Press, 1957.

[4] D. Bertsekas and S. Shreve, *Stochastic Optimal Control, The Discrete Time Case*. Academic Press, New York, 1978.

[5] J. Birge, "Decomposition and partitioning methods for multistage stochastic linear programs," *Operations Research*, vol. 33, 1985, pp. 989–1007.

[6] R. Birge and F. Louveaux, *Introduction to Stochastic Programming*, 2nd. New York: Springer, 2011.

[7] J. F. Bonnans, Z. Cen, and T. Christel, "Energy contracts management by stochastic programming techniques," *Annals of Operations Research*, vol. 200, 2012, pp. 199–222.

[8] J. F. Bonnans and A. Shapiro, *Perturbation Analysis of Optimization Problems*, ser. Springer Series in Operations Research. Springer, 2000.

[9] B. da Costa and V. Leclére, "Dual SDDP for risk-averse multistage stochastic programs," *Operations Research Letters*, vol. 51, 2023, pp. 332–337.

[10] G. Dantzig, "Linear programming under uncertainty," *Management Science*, vol. 1, 1955, pp. 197–206.

[11] L. Ding, S. Ahmed, and A. Shapiro, "A python package for multistage stochastic programming," *Optimization online*, 2019. URL: http://www.optimization-online.org/DB_FILE/2019/05/7199.pdf.

[12] C. J. Donohue and J. R. Birge, "The abridged nested decomposition method for multistage stochastic linear programs with relatively complete recourse," *Algorithmic Operations Research*, vol. 1, no. 1, 2006.

[13] O. Dowson and L. Kapelevich, "SDDP.jl: A Julia Package for Stochastic Dual Dynamic Programming," *Informs Journal on Computing*, vol. 33, 2021, pp. 27–33.

[14] M. Dyer and L. Stougie, "Computational complexity of stochastic programming problems," *Mathematical Programming*, vol. 106, 2006, pp. 423–432.

[15] H. Föllmer and A. Schied, *Stochastic Finance: An Introduction in Discrete Time*, 2nd. Walter de Gruyter, Berlin, 2004.

[16] C. Füllner and S. Rebennack, "Stochastic dual dynamic programming and its variants," *Optimization online*, 2021.

[17] S. Ghadimi, A. Ruszczynski, and M. Wang, "A single timescale stochastic approximation method for nested stochastic optimization," *SIAM Journal on Optimization*, vol. 30, no. 1, 2020, pp. 960–979.

[18] P. Girardeau, V. Leclére, and A. B. Philpott, "On the convergence of decomposition methods for multistage stochastic convex programs," *Mathematics of Operations Research*, vol. 40, 2016, pp. 130–145.

[19] V. Guigues, *Inexact cuts in deterministic and stochastic dual dynamic programming applied to linear optimization problems*, 2018. arXiv: 1801.04243 [math.OC].

[20] V. Guigues and M. Bandarra, "Single cut and multicut sddp with cut selection for multistage stochastic linear programs: Convergence proof and numerical experiments," 2019.

[21] V. Guigues, A. Shapiro, and Y. Cheng, "Duality and sensitivity analysis of multistage linear stochastic programs," *European Journal of Operational Research*, vol. 308, 2023, pp. 752–767.

[22] V. Guigues, A. Shapiro, and Y. Cheng, "Risk-averse stochastic optimal control: An efficiently computable statistical upper bound," *Operations Research Letters*, vol. 51, 2023, pp. 393–400.

[23] G. Hanasusanto, D. Kuhn, and W. Wiesemann, "A comment on computational complexity of stochastic programming problems," *Mathematical Programming*, vol. 159, 2015, pp. 557–569.

[24] M. Hindsberger and A. Philpott, "Resa: A method for solving multistage stochastic linear programs," *Journal of Applied Operational Research*, vol. 6, no. 1, 2014, pp. 2–15.

[25] C. Ju and G. Lan, "Dual dynamic programming for stochastic programs over an infinite horizon," *arXiv*, 2023.

[26] J. Kelley, "The cutting-plane method for solving convex programs," *Journal of the Society for Industrial and Applied Mathematics*, vol. 8, 1960, pp. 703–712.

[27] G. Lan, "An optimal method for stochastic composite optimization," *Mathematical Programming*, vol. 133(1), 2012, pp. 365–397.

[28] G. Lan, "Bundle-level type methods uniformly optimal for smooth and nonsmooth convex optimization," *Mathematical Programming*, vol. 149(1-2), 2015, pp. 1–45.

[29] G. Lan, *First-order and Stochastic Optimization Methods for Machine Learning*. Switzerland AG: Springer Nature, 2020.

[30] G. Lan, "Policy mirror descent for reinforcement learning: Linear convergence, new sampling complexity, and generalized problem classes," 2021.

[31] G. Lan, "Complexity of stochastic dual dynamic programming," *Mathematical Programming*, vol. 191, 2022, pp. 717–754.

[32] G. Lan, "Policy optimization over general state and action spaces," *arXiv preprint arXiv:2211.16715*, 2022.

[33] G. Lan, A. S. Nemirovski, and A. Shapiro, "Validation analysis of mirror descent stochastic approximation method," *Mathematical Programming*, vol. 134, 2012, pp. 425–458.

[34] G. Lan and Z. Zhou, "Dynamic stochastic approximation for multi-stage stochastic optimization," *Mathematical Programming*, vol. 187, 2021, pp. 487–532.

[35] V. Leclére, P. Carpentier, J.-P. Chancelier, A. Lenoir, and F. Pacaud, "Exact converging bounds for Stochastic Dual Dynamic Programming via Fenchel duality.," *SIAM J. Optimization*, vol. 30, 2020, pp. 1223–1250.

[36] C. Lemaréchal, A. Nemirovskii, and Y. Nesterov, "New variants of bundle methods," *Mathematical Programming*, vol. 69, 1995, pp. 111–147.

[37] J. Liang and R. D. Monteiro, "A proximal bundle variant with optimal iteration-complexity for a large range of prox stepsizes," *SIAM Journal on Optimization*, vol. 31, no. 4, 2021, pp. 2955–2986.

[38] N. Löhndorf and A. Shapiro, "Modeling time-dependent randomness in stochastic dual dynamic programming," *European Journal of Operational Research*, vol. 273, 2019, pp. 650–661.

[39] N. Löhndorf, D. Wozabal, and S. Minner, "Optimizing trading decisions for hydro storage systems using approximate dual dynamic programming," *Operational Research*, vol. 61, 2013, pp. 810–823.

[40] V. L. de Matos, A. B. Philpott, and E. Finardi, "Improving the performance of stochastic dual dynamic programming," *Journal of Computational and Applied Mathematics*, vol. 290, 2015, pp. 196–208.

[41] R. Munos, "Error bounds for approximate policy iteration," in *19th International Conference on Machine Learning*, pp. 560–567, 2003.

[42] R. Munos, "Error bounds for approximate value iteration," in *20th national conference on Artificial intelligence*, pp. 1006–1011, 2005.

[43] R. Munos and C. Szepesvári, "Finite-time bounds for fitted value iteration," *Journal of Machine Learning Research*, 2008, pp. 815–857.

[44]  A. S. Nemirovski, A. Juditsky, G. Lan, and A. Shapiro, "Robust stochastic approximation approach to stochastic programming," *SIAM J. Optimization*, vol. 19, 2009, pp. 1574–1609.

[45]  A. S. Nemirovski and D. Yudin, *Problem complexity and method efficiency in optimization*, ser. Wiley-Interscience Series in Discrete Mathematics. John Wiley, XV, 1983.

[46]  Y. Nesterov and A. Nemirovski, *Interior-Point Polynomial Algorithms in Convex Programming*. Philadelphia: SIAM, 1994.

[47]  M. Pereira and L. Pinto, "Multi-stage stochastic optimization applied to energy planning," *Mathematical programming*, vol. 52, no. 1-3, 1991, pp. 359–375.

[48]  G. Pflug, "Some remarks on the value-at-risk and the conditional value-at-risk," in *Probabilistic Constrained Optimization: Methodology and Applications, S. Uryasev (Ed.)* Kluwer Academic Publishers, Norwell, MA, 2000.

[49]  A. Philpott, V. d. Matos, and E. Finardi, "On solving multistage stochastic programs with coherent risk measures," *Operations Research*, vol. 61, 2013, pp. 957–970.

[50]  A. Philpott, F. Wahid, and F. Bonnans, "A mixed integer dynamic approximation scheme," *Mathematical Programming*, vol. 181, 2020, pp. 19–50.

[51]  A. Philpott and V. de Matos, "Dynamic sampling algorithms for multi-stage stochastic programs with risk aversion," *European Journal of Operational Research*, vol. 218, 2012, pp. 470–483.

[52]  A. Philpott, V. de Matos, and E. Finardi, "On solving multistage stochastic programs with coherent risk measures," *Operations Research*, vol. 61, no. 4, 2013, pp. 957–970.

[53]  A. Pichler and A. Shapiro, "Mathematical foundations of distributionally robust multistage optimization," *SIAM J. Optimization*, vol. 31, 2021, pp. 3044–3067.

[54]  B. Polyak, "New stochastic approximation type procedures," *Automat. i Telemekh.*, vol. 7, 1990, pp. 98–107.

[55]  B. Polyak and A. Juditsky, "Acceleration of stochastic approximation by averaging," *SIAM J. Control and Optimization*, vol. 30, 1992, pp. 838–855.

[56] W. Powell, *Approximate Dynamic Programming: Solving the Curses of Dimensionality*, 2nd. New York: John Wiley and Sons, 2011.

[57] H. Robbins and S. Monro, "A stochastic approximation method," *Annals of Mathematical Statistics*, vol. 22, 1951, pp. 400–407.

[58] R. T. Rockafellar and R. J.-B. Wets, "Scenarios and policy aggregation in optimization under uncertainty," *Mathematics of Operations Research*, vol. 16, no. 1, 1991, pp. 119–147.

[59] R. T. Rockafellar, *Conjugate Duality and Optimization*. Philadelphia: Society for Industrial and Applied Mathematics, 1974.

[60] R. Rockafellar and S. Uryasev, "Conditional value-at-risk for general loss distributions," *J. of Banking and Finance*, vol. 26, no. 7, 2002, pp. 1443–1471.

[61] R. Rockafellar, *Convex Analysis*. Princeton University Press, 1970.

[62] A. Ruszczyński and A. Shapiro, "Conditional risk mappings," *Mathematics of Operations Research*, vol. 31, 2006, pp. 544–561.

[63] A. Ruszczyński and A. Shapiro, "Optimization of convex risk functions," *Mathematics of Operations Research*, vol. 31, 2006, pp. 433–452.

[64] A. Shapiro, "On complexity of multistage stochastic programs," *Operations Research Letters*, vol. 34, 2006, pp. 1–8.

[65] A. Shapiro, "Analysis of stochastic dual dynamic programming method," *European Journal of Operational Research*, vol. 209, 2011, pp. 63–72.

[66] A. Shapiro and Y. Cheng, "Central limit theorem and sample complexity of stationary stochastic programs," *Operations Research Letters*, vol. 49, 2021, pp. 676–681.

[67] A. Shapiro and Y. Cheng, "Dual bounds for periodical stochastic programs," *Operations Research*, 2021.

[68] A. Shapiro, D. Dentcheva, and A. Ruszczyński, *Lectures on Stochastic Programming: Modeling and Theory*, third. Philadelphia: SIAM, 2021.

[69] A. Shapiro and L. Ding, "Periodical multistage stochastic programs," *SIAM J. Optimization*, vol. 30, 2020, pp. 2083–2102.

[70] A. Shapiro and A. Nemirovski, "On complexity of stochastic programming problems," in *Continuous Optimization: Current Trends and Applications: Current Trends and Applications*, V. Jeyakumar and A. Rubinov, Eds., Springer, 2005, pp. 111–144.

[71] A. Shapiro, W. Tekaya, J. da Costa, and M. P. Soares, "Risk neutral and risk averse stochastic dual dynamic programming method," *European Journal of Operational Research*, vol. 224, 2013, pp. 375–391.

[72] P. Shinde, I. Kouveliotis-Lysikatos, and M. Amelin, "Multistage stochastic programming for vpp trading in continuous intraday electricity markets," *IEEE Transactions on Sustainable Energy*, vol. 13, 2022, pp. 1037–1048.

[73] M. Wang, E. X. Fang, and H. Liu, "Stochastic compositional gradient descent: Algorithms for minimizing compositions of expected-value functions," *Mathematical Programming*, vol. 161, no. 1-2, 2017, pp. 419–449.

[74] S. Zhang and X. Sun, "Stochastic dual dynamic programming for multistage stochastic mixed-integer nonlinear optimization," *Mathematical Programming*, vol. 196, 2022, pp. 935–985.

[75] Z. Zhang and G. Lan, "Optimal methods for convex nested stochastic composite optimization," *Mathematical Programming*, 2020.

[76] P. Zipkin, *Foundation of inventory management*. McGraw-Hill, 2000.

[77] J. Zou, S. Ahmed, and X. A. Sun, "Stochastic dual dynamic integer programming," *Mathematical Programming*, vol. 175, 2019, pp. 461–502.

Milton Keynes UK
Ingram Content Group UK Ltd.
UKHW020009040624
443552UK00005B/252